Theatrocracy

Theatrocracy is a book about the power of the theatre, how it can affect the people who experience it, and the societies within which it is embedded. It takes as its model the earliest theatrical form we possess complete plays from, the classical Greek theatre of the fifth century BCE, and offers a new approach to understanding how ancient drama operated in performance and became such an influential social, cultural, and political force, inspiring and being influenced by revolutionary developments in political engagement and citizen discourse. Key performative elements of Greek theatre are analyzed from the perspective of the cognitive sciences as embodied, live, enacted events, with new approaches to narrative, space, masks, movement, music, words, emotions, and empathy. This groundbreaking study combines research from the fields of the affective sciences – the study of human emotions – including cognitive theory, neuroscience, psychology, artificial intelligence, psychiatry, and cognitive archaeology, with classical, theatre, and performance studies.

This book revisits what Plato found so unsettling about drama – its ability to produce a *theatrocracy*, a "government" of spectators – and argues that this was not a negative but an essential element of Athenian theatre. It shows that Athenian drama provided a place of alterity where audiences were exposed to different viewpoints and radical perspectives. This perspective was, and is, vital in a freethinking democratic society where people are expected to vote on matters of state. In order to achieve this goal, the theatre offered a dissociative and absorbing experience that enhanced emotionality, deepened understanding, and promoted empathy. There was, and still is, an urgent imperative for theatre.

Peter Meineck is Professor of Classics in the Modern World at New York University, USA. He founded Aquila Theatre in 1991 and has since produced and directed more than 50 professional classical theatre works. He has also directed several National Endowment for the Humanities classics-based public programs, including the Chairman's Special Award-winning Ancient Greeks/Modern Lives and The Warrior Chorus national veteran's program. He has written widely on ancient theatre and its reception, and has published several translations of Greek drama.

Theatrocracy

Greek Drama, Cognition, and
the Imperative for Theatre

Peter Meineck

Routledge
Taylor & Francis Group

LONDON AND NEW YORK

First published 2018 by Routledge

2 Park Square, Milton Park, Abingdon, Oxfordshire OX14 4RN

52 Vanderbilt Avenue, New York, NY 10017

Routledge is an imprint of the Taylor & Francis Group, an informa business

First issued in paperback 2020

British Library Cataloguing-in-Publication Data
A catalogue record for this book is available from the British Library

Library of Congress Cataloging-in-Publication Data
Names: Meineck, Peter, 1967– author.
Title: Theatrocracy : Greek drama, cognition, and the imperative for
    theatre / Peter Meineck.
Description: Abingdon, Oxon ; New York : Routledge, 2017. | Includes
    bibliographical references and index.
Identifiers: LCCN 2017001649 | ISBN 9781138205529 (hardback : alk.
    paper) | ISBN 9781315466576 (ebook)
Subjects: LCSH: Greek drama—History and criticism. | Greek drama—
    Modern presentation. | Theater—Greece—History and criticism.
Classification: LCC PA3131 .M393 2017 | DDC 882/.0109—dc23
LC record available at https://lccn.loc.gov/2017001649

ISBN: 978-1-138-20552-9 (hbk)
ISBN: 978-0-367-59493-0 (pbk)

Typeset in Times New Roman
by Apex CoVantage, LLC

For Desiree, Sofia, and Marina
always on my mind

Contents

Acknowledgements

I would like to thank the many people who have encouraged this work, exchanged ideas, read drafts, offered comments on my talks, and generally have been sympathetic to what is, after all, a relatively new field in classical studies. These include Alan Sommerstein at Nottingham, who encouraged my early adventures in cognitive science; David Konstan, my esteemed colleague at New York University (NYU), who has always been passionately interested in what made the ancients tick; Douglass Cairns, who as director of the History of Distribution Cognition project gathered a remarkable group of people working on cognitive approaches to the ancient world and deepened my understanding of empathy; and Mark Griffiths at Berkeley, who offered excellent notes on ancient music. I also wish to thank Nina Coppolino, editor at the *New England Classical Journal*, and Nicholas Poburko, the editor at Arion, for their help in publishing earlier versions of some of this material. Thanks also to all of those who invited me to talk on this subject and learn from the many questions these events produced, including Tom Habinek at University of Southern California, Walter Scheidel at Stanford, Tom Palaima at University of Texas Austin, Ineke Sluiter at Leiden, Renaud Gagné at Cambridge, Chris Carey at University College London, Emmanuela Bakola at King's College London, Judith Mossman at Nottingham, the Classics faculties at Princeton, Cornell, and Wesleyan, and as the Onassis Fellow at the Classical Association of New England and at the Harvard Center for Hellenic Studies in Washington, DC. I was also fortunate in being able to conduct a number of practical workshops on this subject, several in Greece including at the Michael Cacoyannis Foundation, the Onassis Foundation, and the National Hellenic Research Foundation, at the Association for Theatre in Higher Education Meeting in Los Angeles, and at the "Unpacking the Emotions" conference hosted by NYU's Emotional Brain Institute and Center National de la Recherché Scientifique.

The American Philological Association (now the Society for Classical Studies) in North America allowed me to convene their first panel session on Cognitive Classics in New Orleans in 2015, and I am very grateful to Ineke Sluiter, Garret Fagan, William Short, Jacob Mackey, and Jennifer Deveraux for their participation, encouragement, and fascinating discussions. We met again, along with many others, as this field is rapidly growing, at the Classics and Cognitive

Theory conference at NYU in 2016. I look forward to many more collaborations, exchanges, conferences, and publications to come.

A large part of this book was written at the Bedford Fire Department in Westchester, New York, where I hope I repaid my desk time by going on enough fire and medical calls. My special thanks goes to Jennifer Deveraux for reading and commenting on the entire manuscript, my colleagues in the wonderful classics department at NYU for their continued support, and my wife Desiree for her love and encouragement.

Introduction
Theatre as a mimetic mind

This is a book about the power of the theatre – how it can affect the people who experience it and the societies within which it is embedded. It takes as its model the earliest theatrical form we possess complete plays from, the classical Greek theatre of the fifth century BCE, and uses a new approach to understanding how ancient drama operated in performance and became such an influential social, cultural, and political force, inspiring and being influenced by revolutionary developments in political engagement and citizen discourse. With this in mind, I approach Greek theatre from the perspective of the cognitive sciences as an embodied live-enacted event, and I analyze how certain different performative elements acted upon its audiences to create absorbing narrative action, emotional intensity, intellectual reflection, and strong feelings of empathy. This was the key to the transformative artistic and social power that enabled Greek drama to advance alternate viewpoints and display distinctly different perspectives. In Athens, theatre – an art form associated with cult practice – soon became a major part of Athenian political engagement. This led Aristotle to comment on how the collective audience displayed the best abilities of human judgment, and for Plato to bitterly complain that this theatrical empowerment was allowing for a kind of rule of the masses, what Plato called a *theatrocracy*, the title of this book.[1]

I revisit what Plato found so unsettling about drama – its ability to produce a *theatrocracy*, a "government" of spectators – and argue that this was not a negative but an essential element of Athenian theatre. I hope to show that Athenian drama provided a place of alterity where audiences were exposed to different viewpoints and radical perspectives. This perspective was, and is, vital in a free-thinking democratic society where people are expected to vote on matters of state. In order to achieve this goal, the theatre offered a dissociative and absorbing experience that enhanced emotionality, deepened understanding, and promoted empathy. Paul Cartledge has described the Athenian theatre as an essential part of the learning process of the Athenian citizen "to be an active participant in self-government by mass meeting and open debate between peers."[2] My aim is to explore how the theatre did this by analyzing several important experiential elements from a variety of new perspectives: narrative, environment, masks, movement, music, and lyrics.

Plato was incensed that ordinary citizens felt that they had the right to express a political opinion, and certainly, theatre can manipulate emotions, misdirect, push certain agendas, and tacitly support particular political, religious, or cultural values. Yet, even with the advent of the spectacular mimetic attributes of film and television, which both employ many of the same performative devices we first find in Greek drama, theatre remains a part of our cultural landscape, with more than 12.9 million people visiting Broadway shows in New York City in 2015 alone.[3] But in order to better understand what made ancient Greek drama such a powerful cultural force, it is important to emphasize the differences between the performance of these works in Athens 2,500 years ago and the experience of theatre today. Hence, this book focuses on Greek drama *in* and *as performance*, and seeks to add to our knowledge of the art form by asking how the plays would have affected the audience who experienced them as live performative events.

Most past studies on ancient drama have tended to view the art form either through the prism of the theatrical aesthetics of their own day or primarily as works of literature. A cognitive approach can help define the fundamental differences between ancient Greek drama and the theatre of the 20th and 21st centuries and offer us the means to explore the experiential elements of what made Greek drama so distinctive. This is important if we want to more fully understand how Greek drama functioned, was originally received, and became as popular and influential as it did. With this in mind, I explore these differences and examine some of the key experiential elements of Greek drama: (a) narrative as predictive stimulus; (b) open-air space as a mind-altering property; (c) the mask as an effective emotional material anchor; (d) movement, gestures, and chorality as powerful collective kinesthetic communicators; (e) the properties of music for cognitive absorption; and (f) the dissociative elements of dramatic speech and song. These are placed within a social and political context as part of the empathy-generating *gesamtkunstwerk* of Greek drama. The scope of this book means it is not possible to discuss every element of Greek theatre production in detail, but there has recently been good work on props from cognitive perspectives from Chaston and Mueller, and I hope this study will help encourage more of these types of approaches.[4] To structure this study, I revisit Aristotle's six constituent parts of tragedy: *muthos* (narrative), *opsis* (visuality), *ethos* (character), *dianoia* (intention), *melos* (music), and *lexis* (words), offering new experiential perspectives on familiar poetic themes. Consequently, although comedy, satyr drama, and dithyramb are all referenced, the focus will be on the performance of fifth-century Athenian tragedy.

This book has four main aims: first, to explore what was so distinctive about classical Athenian drama from an experiential perspective. Second, to demonstrate how the cognitive sciences can help reveal how the ancient theatre operated in performance and also fulfilled a wider cultural role in the creation of social empathy and political discourse. Third, to show how the cognitive sciences can be of great help to those interested in understanding more about the ancient world. In this respect, I hope this book helps develop new models of inquiry that can prove fruitful when applied to many aspects of the study of antiquity. Fourth, in exploring some of the fundamentals of Greek drama, to highlight why there is an imperative for theatre in democratic Athens and by reflection how theatre can still be of great cultural value today.

The genesis of this study was to ask a most basic question: why did Athenian drama grow to become such a major cultural influence, one that still reverberates around the world today? I argue that it is because of the way in which it originally affected its audiences, not only as they sat in the theatre but also as it impacted their social and political lives. Though we have come to know these plays via the surviving texts, their creation and survival first depended on them being enacted before a live audience in Athens in the fifth century BCE. Therefore, a close study of the affective and experiential aspects of Greek drama is essential and has not been undertaken before from a cognitive perspective. However, there have been some important steps in this direction: in 1980, W. B. Stanford produced a small but influential work, *Greek Tragedy and The Emotions*, that was interested in understanding the emotional power of Greek drama; in 2002, Rush Rehm's *The Play of Space* applied the environmental theories of James Gibson to the performative dynamics of the Greek theatre; in 2010, Felix Budelmann and Pat Easterling began to explore the potential for cognitive approaches to ancient drama; and Douglass Cairns has been profitably exploring Greek literature, including drama, from an emotional perspective for several years now.[5] In addition to applying cognitive theories to ancient drama, this book will also utilize research from the fields of neuroscience, psychology, robotics, and artificial intelligence, all areas that are interested in the affective and experiential parts of human cognition. Classicists, archaeologists, and ancient historians have been studying the cultures of the ancient Greeks since the Renaissance, albeit refracted via their own various social milieus. If we accept that human cognition is bio-cultural and that cultures are created by human minds, then it would only be prudent to examine the biological side of human cognition when considering an ancient culture. What then can cognitive studies and neuroscience contribute to our understanding of the ancient world? Thomas Habinek provides a lucid answer:

> It's not that neuroscience provides definitive answers; rather, by articulating a model of throught and action radically different from those taken for granted by most scholars, neuroscience defamiliarizes the ancient material, opening up new horizons of understanding, much as comparative ethnography and critical theory have done for previous generations of classicists. Neuroscience teaches us very little about the essential nature of the human organism, except that it is constantly changing through "inhabited interaction" with the material universe. But it gives us excellent tools for understanding the constraints upon and characteristics of such interaction. In that sense, it can't help getting inside the heads of humanists, metastasizing into our disciplinary bodies.[6]

One benefit of this interdisciplinary approach is that the clinical aspects of neuroscience allow us to distance ourselves slightly from our own cultural biases when we examine aspects of antiquity. Another is that by thinking with neuroscience, we might approach material from a different perspective and form new conclusions. This is certainly a new methodology for classical studies, but several scholars have recently applied neuroscience research and cognitive theory to various facets of the theatre arts with a good deal of success. For example, Rhonda Blair has used neuroscience research to develop a new framework to understand and describe the discipline of acting for the theatre; Evelyn Tribble has explored the mnemonic features of Shakespeare's poetry through the theory of distributed cognition, encompassing the designation of roles, plotlines, spatial considerations, and Elizabethan playbooks; and Naomi Rokotnitz has examined embodiment in drama as a means of changing the somatic identity of actors and audiences and has applied her theories to plays by Shakespeare, Stoppard, Wertenbaker, and Kaufman.[7] Additionally, Bruce McConachie has produced a groundbreaking cognitive study of theatre spectatorship, and John Lutterbie has applied dynamic systems theory and embodied cognition to different theories of actor training.[8] A recent fruitful interdisciplinary collaboration between Shakespeare scholar Evelyn Tribble and cognitive scientist John Sutton is a proposal to approach historical theatre works from the perspective of a cognitive ecology, which facilitates a systems-level analysis of theatre.[9] The authors write: "this model of cognitive ecology would posit that a complex human activity such as theater must be understood across the entire system, which includes such elements as neural and psychological mechanisms underpinning the task dynamics." These include the body movements and gestures of the actors, the physical environment including the actor/audience relationship, theatre technologies, economies, social and cultural impacts, prevailing aesthetic preferences, and what the authors describe as "cognitive artifacts" such as parts, plots, and playbooks.

How then might we profitably apply some of the same cognitive approaches to ancient drama? We have little to go on. We possess only seven complete plays of Aeschylus, seven of Sophocles, eighteen of Euripides (not including *Rhesus*, which seems to be a fourth-century play), eleven of Aristophanes, and only one complete play of Menander. Even with the other fragments of these and other playwrights, what we have is a small fraction of the total theatrical output of the classical Athenian theatre. Also, the texts do not tell the entire story; they are at best, as Philip Auslander has written, "blueprints for performance."[10] Even the oldest texts and fragments we possess, found on mummy cartonnage and in ancient rubbish dumps in the remains of the Greek communities of Egypt, are still not authorial play scripts but most probably scholastic texts or copies from private libraries. They have no stage directions, and apart from later comments added by the so-called scholiasts, no indication of stagecraft, costume, masks, dance steps, music, or movements. Instead, we are left to make educated inferences from what we can glean from the surviving and sometimes corrupt texts and what we know of the staging conditions of the original play. This last aspect also needs revisiting, as recent archaeological surveys of the site of the Theatre of Dionysos in Athens have questioned established ideas about the size, form, and material of the classical Greek theatre, issues that will be described and addressed in this book.[11]

A cognitive approach allows us to re-examine the textual, material, and anecdotal evidence of the Greek theatre and place it into an experiential context. This assumes the basic premise that theatre is created to be enacted and consumed by an audience, and in order to be understood it must communicate its intentions and emotions effectively. Therefore, a culture's theatre is a kind of mimetic mind, an artificial construct that mirrors, amplifies, and projects the cognitive regime of the people who have come to experience it. This understanding of the experience of Greek drama stands at the heart of this study.

I have been involved in making and teaching about theatre for nearly 30 years, beginning in 1987 with a student production of Aeschylus's *Agamemnon* at University College London's Bloomsbury Theatre. Since then I have worked as a technician, production manager, producer, director, translator, and writer and founded Aquila Theatre in London in 1991. With Aquila I have been lucky enough to stage classical plays all over the world, at venues as diverse as the ancient stadium at Delphi, Carnegie Hall, the Lincoln Center, the Brooklyn Academy of Music, the Assembly Rooms in Edinburgh, various theatres in New York, London, and Athens, at performing arts centers and festivals throughout North America and Europe, and at two different performances at the White House. In my parallel career as an academic, I have sought to fuse my knowledge and experience of the practical act of making theatre with a scholarly approach to Greek drama. At all times I have been most interested in how Greek theatre worked in performance and approached ancient texts and evidence with a practitioner's eye, albeit one from a different time and culture. Nevertheless, I have always felt an affinity with ancient drama, ever since my Professor of Greek at UCL, Pat Easterling, took one look at me and said, "you should study Aeschylus, he was a soldier like you" (I was serving in the Royal Marines Reserves at the time). One profound moment in my professional career was a time when an older, much-respected actor I once worked with shook my hand and said "now you are only a handshake away from Shakespeare," emphasizing our proximity to the English theatre of 400 years ago. It is nearly 2,500 years to Aeschylus, and there is no unbroken performance tradition, no connective "handshake," but in all my theatrical travels around the world I have found commonalities and affinities among the performance artists of different cultures that also resonate within the ancient works. I hope then that I can add something to the rich corpus of work by classicists on the ancient Greek theatre, and at the same time offer a slightly different perspective, one that is most interested in theatre as a live art form and asks not only *what* Greek theatre represented but *how* it actually worked in practice.

It is certainly not new to exclaim the cultural importance and influence of Greek drama. What is new in this book is the way in which I am approaching the evidence for the ancient stage to try to understand its effect upon its original audience. Though we are able to read many Greek plays today, the act of reading a play is cognitively distinct from watching, listening, sensing, and experiencing the same play performed live. I am most interested in this multisensory experience of theatre. Here I am inspired by the field of cognitive archaeology, which for the past two decades has been interested in how the material record of ancient cultures reflects ancient thought processes, and how environment and tool use changed human minds. Malafouris and Renfrew have called this "the cognitive life of things" after Ajurn Appadurai's seminal work *The Social Life of Things*, which explored how material objects have agency and the biological dimension of artifacts.[12] What I find so dynamic about this approach is the basic idea that an ancient artifact is not just an aesthetic object or means of dating, but the actual remnant of an ancient thought process. I think we can view the surviving texts and material evidence for ancient drama in much the same way and apply cognitive theory, neuroscience, and psychological research to better understand the experiential elements of ancient drama within their original context.

Distributed cognition

Renfrew, Malafouris, and other cognitive archaeologists seek to "look beyond the brain itself and emphasize the social and cultural context" and cite the work of Andy Clark, who asserts that "human engagement with the material world plays a central role."[13] In this study I draw on these theories of distributed cognition, which posit that the human mind is not situated only in the brain, but is extended via the body out into the environment, in what Andy Clark has described as a "constant cognitive feedback loop."[14] Various theoretical viewpoints broadly agree with the premise that the mind is not "brainbound," and they have been grouped together as "4E theory" – that the mind is embodied, embedded, enacted, and extended. These theories of distributed cognition can be valuable tools for comprehending how theatre works and have been profitably applied in recent theatre studies by the scholars mentioned previously.

If we accept the premise that human cognition is extended, then we can start to discern what Edwin Hutchins termed "cognition in the wild," basically, material elements of cognitive scaffolding that we use to help manage often complex cognitive tasks.[15] Hutchins famously demonstrated the need for air traffic controllers to situate the huge mental task of incoming, taxiing, and outgoing planes on simple paper slips, the manipulation of which proved essential for effective detailed management, despite the availability of powerful computer tools. Malafouris posits that "the content of a mental state is in part determined by elements of the external world, and thus human cognitive skills cannot be studied independent of the external environment."[16] Extended cognition must therefore include exchanges among people, artifacts, environment, and time. Malafouris uses the example of Mycenaean Linear B tablets and goes beyond the deciphering and translation of the writing system to emphasize the human and material interactions. In this way, what Malafouris calls Material Engagement Theory helps us learn more about ancient labor, social practices, and the communicative pathways, both verbal and textual, that indicate informational exchange in both existing and emergent cultural practices. He describes this methodology as

> a shift from the micro level of semantics to the macro level of practice – Linear B is no longer seen as a disembodied abstract code; now it is seen as a situated

technology . . . encompassing reciprocal and culturally orchestrated interactions among humans, situations, tools and space.[17]

These kind of cognitive connections among brain, body, material objects, and the environment are mirrored in the very act of creating theatre, which is nothing less than the making of a mimetic on-stage mind. For theatre to function it needs to reflect, heighten, and distribute the cognitive mechanisms of its audience, and therefore we should also be able to learn something of the cognitive regime of the culture within which that particular theatre tradition operated. This has implications for understanding ancient minds, particularly in light of recent advances in the fields of cultural neuroscience and epigenetics, which have shown how culture, environment, and emotional experience deeply affect human cognition.

Whereas I accept that the machinery of the mind is extended out into the environment, that human minds interact with each other, and that materiality plays a vital role in human cognition, these positions should not, however, negate the role of the brain in these processes. Theories of distributed cognition have been attractive to scholars in the humanities, as they can sometimes seem to offer a rejection of more Cartesian computational approaches and neuroscience in general. For example, Alva Noë has stated that neuroscience as a discipline is still quite underdeveloped and compares it to a teenager in the grip of its own technology.[18] To be sure, there is a popular view of neuroscience as deterministic and universalist, a field that seems to some to threaten the very existence of the humanities, but this is not the case. Neuroscience is a broad and diverse field, which I hope to show has much to offer those who study the humanities. Furthermore, many prominent neuroscientists also hold that human cognition is at least partially, if not fully, extended and distributed.[19] For example, Antonio Damasio has stated that "it is not only the separation between mind and brain that is mythical: the separation between mind and body is probably just as fictional. The mind is embedded in the full sense of the term, not just embrained."[20]

Another relatively new area of cognitive science that will be utilized in this study is the area of cultural neuroscience, which has recently exploded the idea of the universal human mind and shown how much culture both shapes cognition and continues to act upon the brain, which remains plastic and mutable throughout a person's lifetime.[21] Theories of distributed cognition actually go quite far in reconciling this artificial biological and cultural division by eradicating it altogether and situating the physical brain firmly within the culture within which it constantly learns and responds. For example, Miranda Anderson has successfully applied distributed cognition to Renaissance literature and proposes that the paradigm of the extended mind suggest a means of negotiation:

> The social constructionist models that pervade literary studies can be argued to have a physical basis (as our ability to be constructed by sociocultural forces relates to neurological plasticity) at the same time as human extendedness and adaptability (to cultural, physical or linguistic variables) tempers any notion of universals that might be attempted.[22]

If the basic theory of distributed cognition is correct, as I believe it is, in holding that the human mind is extended in a feedback loop between brain, body, environment, and back again, then those models that fail to fully incorporate the internal mechanisms of the brain as part of the entire cognitive process should perhaps be as suspect as strict Cartesian computational stances that equally fail to accommodate the mind's embedded bodily and environmental cognitive scaffolding.

Prediction

What we need then is a fully comprehensive theory of human cognition that might successfully merge the computational mechanisms of the brain as observed by neuroscience with the vital distributed scaffoldings with which we operate in the world and the culture that we act in and upon. Enter Andy Clark, the philosopher whose 1998 paper "The Extended Mind," co-authored with David Chalmers, has had so much influence on those working in distributed cognition. In a groundbreaking 2013 paper and his subsequent 2015 book, Clark has sought to reconcile computational models with theories of distributed cognition, including his own, under a broad theory he names "predictive processing."[23] Clark cites this as a "real clue" to the mystery of human cognition, one that won't solve all the conundrums of consciousness, emotions, and intelligent action, but that can be an "umbrella under which to consider (and in some cases rediscover) many of the previous clues."[24] Clark sums up prediction as the means by which we "deal rapidly and fluently with an uncertain and noisy world." In this way, the brain uses top-down processing to quickly deal with incoming sensory data, using error messages to make predictions as to what is being experienced. Clark's theory incorporates both the bottom-up sensorial information conveyed by ears, eyes, tongues, noses, skin, proprioception (the sense of the body in space), and interoception (visceral states such as pain and hunger) with the top-down "learned information" stored in the brain. These two concurrent streams meet in a constant informational exchange that can rapidly yield sense-making predictions, what Clark calls a kind of "bootstrap heaven." The home for this "inner prediction engine" is described by Clark as a "mobile embodied agent located in multiple empowering webs of material and social structure."

Clark posits that predictive processing can "offer new tools for thinking about the moment by moment orchestration of neural, bodily and environmental forces into effective transient problem-solving solutions . . . an action orientated engagement machine, an enabling node in patterns of dense reciprocal exchange binding body brain and world."[25] Hence, predictive processing goes some way in reconciling the perceived epistemological rift between neuroscience and distributed cognition. But my aim is not to verify or disprove Clark's theory or any of the others I apply in the course of this study. Rather, it is to apply them profitably to certain aspects of what we know of the ancient theatre in the hope that we may come to some new conclusions, and to be emboldened to ask fundamental questions to better understand how ancient drama worked as performance. I examine predictive processing in more detail in Chapter One and relate it directly to several of Aristotle's concepts of dramatic narrative.

Emotions

We have no contemporary critical responses to the plays and no related writings by the playwrights or performers. What we do possess are the faint glimpses of the response to drama found in other texts from the fifth and fourth centuries, and they are all concerned with the extreme emotionality of the theatre. For example, Herodotus tells us that the early tragedian Phrynichus was fined for "causing the audience to fall into grief" for his play *The Sack of Miletus* (6.21.10); the Sophist Gorgias wrote that performed poetry (drama) "forces its hearers to shudder with terror, shed tears of pity, and yearn with sad longing" (*Helen* 9); Xenophon noted how the actor Callippides could move an audience to tears (*Symposium* 3.11); and Isocrates, Plato, and Aristotle all described drama as having the power to "move the soul" (*psychagogia*).[26]

These kinds of marked emotional responses are why in *Laws* Plato described fourth-century Athens as a *theatrocracy* (3.700–701) and complained of the crowd being swayed politically by their emotions because of the theatre's influence. Around this time, Athenian theatre had grown hugely influential. A new, massive, 16,000-seat stone theatre was constructed on the site of the older, smaller, wooden theatre on the southeast slope of the Acropolis, and a *theoric* fund was established, giving poorer Athenians the means to attend. In Book 10 of the *Republic*, Plato has his Athenian argue that theatre should be heavily regulated in the ideal city, and only plays that reflected what he thought were "good" values should be permitted. Plato knew the pervasive power of drama and even has his Socrates complain in the *Apology* that the jurors of the large citizen law court judged him on the basis of his comic portrayal in Aristophanes' *Clouds*.[27] Aristotle was less severe in his judgment of drama, although in *Poetics* he does seem to offer a somewhat nostalgic view of a classical theatre that he never experienced first-hand. For much of the late twentieth century, Aristotle was spurned by theatre scholars as offering only a literary view of drama,[28] but Aristotle has much of value to say on the theatre, though it is surely correct that his *Poetics* should not be held up as any kind of definitive manual for drama. Rather, it is an exploration of the idea that the performance of poetry is a creative art form made up of constituent parts.[29] Theatre-making is certainly that, as anyone who has worked through a rehearsal process will know, but where I part company with Aristotle is with the notion that theatre is only a craft (*techne*) that can be completely explained by dissection. Yes, theatre is made by skill, organization, teamwork, and technique, but to work it also needs talent, inspiration, passion, and mystery, things that are ineffable and enigmatic. Zeitgeist, fashion, politics, publicity, unintentional tipping points, and celebrity all play their part in helping to make a show successful. Although I am taking a scientific and theoretical approach to drama here, I have tried not to forget the importance of the uncanny and the ineffable when it comes to experiencing theatre. There is certainly validity in Plato's idea of the inspired performer we find in *Ion*, and although theatre-making is about making sense, some of theatre's finest moments are non-sensical and unintelligible and operate in the realm of the emotions – not the critical or analytical mind.

An intense experience *moves* us and is often manifested physically by an affective state – we say we feel emotions, and indeed we do. Such feelings can involuntarily cause physical changes in that they affect our pulse, the temperature of our skin, our blood pressure, and rate of breathing; we feel chills, shudder, recoil, and gasp; we cry, laugh, shift awkwardly, and cover our eyes. In the theatre, when we sit together with many of our fellow humans, all focused on the same representation of action, the resulting feelings can be greatly magnified. When we watch drama, we may not always feel the same emotional states presented by the actors, but we frequently take a personal position that can be identified as emotional. Even if we are bored, that boredom will be made manifest in an emotional response, like frustration or even anger, and we reflect those emotions bodily and describe them in embodied terms.

But what do we mean when we talk about emotions? What are emotions exactly, and are the emotions we know today the same as those experienced by people from another culture in the distant past? A specific problem we face in embarking on such a study of historical behaviors is that there currently exists no general consensus on what we actually mean by the term "emotion." Archaeologist Sara Tarlow has put this down to "how we use and intend emotional language" as the reason for the "failed communication between emotional scholars."[30] The classicist G.E.R. Lloyd has also asked a series of important questions about the way we apply the study of emotions to work on the ancient world, among them: does the English language provide an adequate terminology for describing emotions? Are there certain basic emotions, and if so, what are they? Are emotions distinct or do they form a continuum and blend into each other? Are emotions linked to moralities?[31]

The study of human emotions has a long history stretching back to Plato, Xenophon, and Aristotle, who all sought to categorize and explain affective states.[32] In the modern era, Darwin turned his attention to embodied emotions in his seminal study, *The Expression of Emotions in Man and Animals* (1872). This work is still highly influential today in the research of the so-called neo-Darwinists, such as Paul Ekman. Ekman's theory of "basic emotions" posits that a finite number of reductive affective states can be visually identified across different cultures. This is one of the most debated positions in affective science today and will be discussed in more detail in Chapter Three. Also, in the late 19th century, William James and Carl Lange simultaneously proposed theories of affective states, suggesting that emotions are physiological reactions to external events, the processing of which is dependent on cognitive interpretation, for example, "I tremble and so I feel fear." James's theory is an early articulation of what has become known as the embodiment theory of emotions, a bio-cultural blend of instinct and interpretation.[33] Today, most theories of emotions fall somewhere between two distinct theoretical positions: "constructionists" or "cultural relativists" posit that emotional states are learned products of culture, whereas the "universalists" of the psychological or biological school propose that at least some emotions are universal across humans and a product of the evolutionary process. My position is that what we commonly describe as emotions are actually both products of human evolution and refinements of culture, as by adopting the theoretical position of distributed cognition this kind of binary distinction becomes moot. There is also a schism between those who view emotions as resulting from judgment ("appraisal theory") and others who surmise that emotions can be far more instantaneous. For example, if one suddenly falls, fear arises before the mind has had any time to form an apparaisal of what just happened; this is known as "embodiment theory." The philosopher Jesse Prinz sets out the two positions succinctly in his 2012 book *Beyond Human Nature*; he concluded that there is no convincing evidence that we need appraisal judgments to distinguish different emotions and that "the embodiment theory is probably right."[34] However, I also think that emotions can arise from both embodiment and appraisal and that emotional responses are multisensory, complex, fluid, and mutable.

If emotions are affective expressions of both embodied experiences and inference, then evidence of emotional responses in ancient Greek culture can help us to understand more about the function and reception of ancient drama. However, traditional forms of understanding the ancient world, primarily via textual analysis, can be problematic when considering ancient emotions. Angelos Chaniotis has pointed out that the basic physical elements of emotional experience "do not exist in the study of written sources," especially when we are "dealing with human beings who died twenty centuries ago,"[35] and yet in evolutionary terms there is no biological difference between us and the ancient Greeks. While we do have images from the material cultural evidence, Douglas Cairns has noted that "in spite of the wealth of visual representations of the human body that have survived from the ancient world, we have no means of evaluating the messages that these representations convey . . . that is wholly independent of ancient textual evidence" and then we run into "problems of ethnocentric bias that arise when we use our own terminology to describe the emotions of another culture."[36] Cairns is right to point out the question of cultural bias, but this is also true of any kind of interpretation of ancient material, textual or material. Also, if some "basic" biological commonalities in emotional expression are found across cultures, then we should be able to identify them in ancient representations and understand something about how they were received within that particular culture. On this last point, David Konstan's seminal study of the emotions of the ancient Greeks delineates "significant differences" between ancient emotions and our own, the recognition of which is essential for understanding Greek literature and culture. He does concede, however, that basic universal biological "affects" may exist at a deeper level.[37]

A solution to this problem of ethnocentric bias has been proposed by the historian Barbara Rosenwein, who suggested that the affective information gleaned from other cultures, particularly those of the past, be assessed by collating information about the emotional regime of that society, or what she has called "emotional communities."[38] Rosenwein's approach includes gathering the source material for the group in question, problematizing their emotional terminology, consulting the theorists of that period, weighing emotional attitudes to assess the relative value placed on each emotion, and then going further and looking for emotions in "silences," metaphors, and ironies; considering the social role they play; and then tracing changes in attitudes to emotions over time.

At first sight, Rosenwein's methodology in developing a history of emotions seems entirely constructivist, yet she writes that "a history of emotions must not deny the biological substratum of emotions, since it is clear that they are embodied on both the body and the brain." For Rosenwein, the differences between constructionist and biological approaches are not inseparable. She suggests that any history of emotions must address the distinctive characteristics of the society under scrutiny, and she adds, "even bodies (and brains) are shaped by culture."[39]

My position is also bio-cultural in that I accept that all humans have shared certain physical and chemical biological commonalities for the last 80,000 years or so.[40] Additionally, I agree that an evolutionary process that developed to respond to environmental stimuli has honed human affective states. But this does not mean that the human mind is fixed and universal across cultures. On the contrary, human biology is highly plastic – for example, the nascent field of epigenetics has shown how DNA can be altered in as little as one generation in response to extreme environmental factors such as stress.[41] Even within one lifetime, brain networks are adaptable, and individuals can reorder existing brain processes while learning a language, playing a musical instrument, mastering a sport, and so on.[42] Yet, our shared human biology does make many elements of our lives universal, whether we like it or not. For example, it may be culture and lifestyle that produce a diet that leads to a blocked artery, but all of us have an increased chance of survival if we receive chest compressions and the heart-stopping reset of a defibrillator, whether we live in London or Lima, regardless of the ethnic group from which we hail. Likewise, affective states and perceptual processes are both biological and cultural, and the two are interconnected, which can be observed in the way in which certain cultural practices developed. For example, in Chapter Two I take a biological view of the Theatre of Dionysos and explore the way in which the open-air environment of the ancient theatre affected the neurochemistry of the people who gathered there and helped promote abstract thought.[43]

I also take the position that human cultures are the embodied expressions of shared human minds responding to basic biological needs and environmental stimuli in a constant cognitive feedback loop. Survival, food production, shelter, group dynamics, reproduction, and safety are just a few of the basic cultural factors that mitigate cognition and call for neuroplasticity. In this respect, differences in culture stem from environmental disparities and differing cognitive solutions to the same basic underlying survival needs. Human culture is the manifestation of the extended and distributed social mind – put simply, minds make culture and culture makes minds. Classicist Thomas Habinek has expressed a similar position: "it is human nature to construct cultural diversity, and what we call culture alters the biological 'nature' of both individuals and the species."[44] Human biology and human culture are therefore inextricably linked – we are all bio-cultural beings, and we share a basic biology with the ancient Greeks; it is our cultures that are different, and they are manifestations of distributed minds.

The particular subject of Greek tragedy and the emotions was taken up in the early 1980s by classicist W. B. Stanford, who categorized emotionalism in the theatre into three broad categories: (1) the mood of the audience, (2) the physical and psychological conditions of the performance, and (3) the nature of the plays being performed.[45] Stanford emphasized the connections between the theatre and emotional responses but framed tragedy as a "civilizing" development out of an older orgiastic cult of Dionysos. In so doing, he negated some of its more ineffable embodied affective elements, by presenting the idea of the development of more "rational" drama. All we have to do is look again at works such as Euripides' *Bacchae*, produced at the end of the fifth century and therefore, according to Stanford's theory, the most "rational" Greek play we should have, to realize that the Greeks were still fascinated with the ineffable and irrational aspects of their theatre encapsulated by this most uncanny of plays. Nevertheless, Stanford's compact study did set out a useful taxonomy of emotional expressions in Greek drama and examined song, noises, cries and silences, music and the spoken word, and certain visual aspects, and I will explore some of these from a cognitive perspective in Chapter Six. One major criticism I have of Stanford's study is that it reflects the prevailing view that the dramatic mask was a fixed, immovable visage that "distanced" the audience in a kind of Brechtian emotional disengagement. Neuroscience studies on facial processing can help us understand that this was not at all the case and that the mask needs to be seriously reappraised as one of the main conveyers of embodiment and emotionality in Greek drama. This will be explored in detail in Chapter Three.

A bio-cultural approach to an important aspect of an ancient culture has been recently advanced by Garrett Fagan, who has applied studies in modern crowd psychology to analyze Roman responses to organized spectacles of violence. He terms his approach "psychobiological" and writes that "an interdependence between contextual stimulus and psychological propensity shapes behavior."[46] Fagan makes the important point that if there were not basic human universal psychological functions across different cultures, then "alien societies ought to remain virtually impenetrable to an outsider," and with this in mind, "it is possible for modern minds to comprehend, analyze and even empathize with the actions of people in other historical eras."[47] This is certainly true of ancient drama – while modern audiences may not grasp the significance of certain ritual actions, religious beliefs, or cultural practice, they can still be moved by the incidents that arise from them.[48]

An important bio-cultural theory of affective states has also been used by Robert Kaster to examine Roman texts for affective information. The psychologist Silvan Tomkins proposed "Affect Theory" – that there are nine hard-wired emotional responses – which he described as "the biological portion of emotion" mediated by socio-cultural factors.[49] Tomkins then proposed "script theory," which posited that when a human is affected by an emotion, the cognitive responses following the basic biological reaction form a kind of behavioral "script," which is also informed by culture. Robert Kaster has applied script theory to Roman texts, making it clear that no Latin emotional term maps perfectly onto a corresponding English one. Kaster's approach is to analyze the affect "from penetration of a phenomenon, through evaluation and response" in order to understand the emotion "through Roman eyes and not through the filter of our own sensibilities."[50]

In his stimulating study, Robert Kaster states "we have only the Roman's words" and makes this a justification for focusing on words alone.[51] However, Chaniotis has suggested that although clearly useful, a lexical approach alone cannot be the "sole methodology for a detailed investigation of the emotional concepts of another culture."[52] Indeed, we have much more than only texts, including the remains of the material culture and the wealth of cognitive information that can be gleaned from vase paintings, sculpture, architecture, sanctuary layouts, town planning, and environmental relationships. As Chaniotis has pointed out, while historians of the ancient world cannot directly study the neurobiological processes of ancient people, they do have access to much of the external stimuli that generated these people's emotions. Distributed cognition asserts that these external stimuli are as much a part of the human mind as the physical brain itself, and that by applying research from the neuro- and affective sciences, we can significantly add to our arsenal of resources for understanding the ancient world. Luther Martin has called this kind of trans-historical approach "a cognitive historiography," suggesting that it can provide correctives to traditional historiographical tools:

> It can do so by identifying and explaining data that have been produced by ordinary processes of human cognition but that have otherwise been neglected in favor of more explicit forms of evidence that historians have, for one reason or another, come to privilege – principally texts, which are themselves, of course, constrained products of human minds to be explained rather than unembellished reservoirs of historical facts. Cognitive theories can contribute insights into how and why some representations of historical behaviors emerged, were favored and remembered, but not others that may have been historically, culturally or cognitively possible or present.[53]

Comparative social modeling

Another reason why a close study of the experience of Greek drama is valuable is that it shares many of the basic performative forms of several other observable cultural theatrical traditions. These include the use of masking, which is also found in Japanese, Chinese, Indonesian, Indian, and many far older shamanistic performance traditions throughout the world. Additionally, Greek theatre was dance and song drama and relied heavily on gestures and movement, both collective and individual, to convey the emotionality of its narratives. We can observe a similar attention to movement in the Indian *Nātyaśāstra* texts and the living traditions of Kathakali, Noh theatre, and Indonesian Topeng theatre. In the fields of cognitive archaeology and ethnography, this practice of comparative social modeling has been used to learn more about ancient artifacts and the cultures that produced them. This is achieved by comparing what we know of a culture that no longer exists with basic empirical similarities found in a similar one that still does.[54] This has been termed the general comparative approach and strives to reconstruct the perceptual information that is formed by both the surrounding environment and concepts stored in working memory, known as "indirect percepts." Such studies have been utilized to compare the Paleolithic rock painting of animals with similar paintings made by existing hunter-gatherer groups in Australia. The conceptual information that produced the paintings in the existing culture are then analyzed to suggest ethnographic correlates with the Paleolithic group. The basic premise of the general comparative approach is that human groups that share certain basic societal characteristics will also share some similar perceptual functions.[55]

A fundamental question for this study is whether we usefully apply such methods to the study of the cultures of antiquity. One place where this debate has recently raged has been in the comparison of combat trauma between modern clinical studies and ancient accounts exemplified by the works of the psychiatrist Jonathan Shay and the ancient historian and combat veteran Lawrence Tritle.[56] There are certainly ethnographic correlates of value here between the ancient Greek hoplite and the modern Western infantryman both in terms of environment and societal percepts. These can be concepts of nationalism, interpersonal relationships forged by military training, the shared experience of combat, democratic participation, separation from family group, loyalty conflicts between family and state, and so on. By this measure, the modern combat veteran can bring us somewhat closer to the Athenian audience, all the members of which had firsthand experience of war, than to persons who have never been affected by the emotionally heightened experience of combat. However, while comparing the psychologies of two distinct cultures may be revealing, we need to be aware of the underlying cultural differences and be very cautious not to impose our own social biases. Shay and Tritle have both noted fascinating parallels between modern symptoms of posttraumatic stress and descriptions found in Greek literature. Yet, this work has been challenged as "universalistic" and even viewed as part of a wider cultural phenomenon of the globalization of American/Western notions of trauma therapy.[57]

This complex issue is highlighted by Ethan Watters, who uses the Western response to the Sri Lankan tsunami in 2004 as an example of a possibly serious misplacement of cross-cultural psychology's good intentions. Watters reports how thousands of emergency therapists and "parachute researchers" poured into the country in the weeks after the enormous natural disaster and spent little or no time attempting to understand the culture or language of the people they were trying to help. It is a relatively new idea that immediate psychological therapy involving direct confrontation of the traumatic event could be the equivalent of emergency medical triage for physical injuries. This therapy has proved problematic when applied to a different cultural group that has its own specific means of dealing with traumatic events. American posttraumatic stress disorder (PTSD) therapies tend to focus on individual response, a separation of the traumatized person from their usual routines and surroundings, and an immediate confrontation of the memories of the traumatic event. This form of treatment clashed with certain deeply held beliefs in Sri Lanka, which comprises a number of different cultural groups that tend to rely on existing social and familial networks to provide therapeutic recovery. Aid workers were shocked when many people taken to refugee camps chose to quickly leave and return home to their devastated villages, or that children expressed a desire to go back to school. Watters reports that many of the Western trauma specialists, unaware of local cultural norms and the importance of social ties, put this down to a mass expression of "the denial phase" of traumatic response.

What therapists trained in Western medicine failed to appreciate is that although many Sri Lankans were clearly traumatized by the tsunami, many displayed symptoms and behavioral patterns not in line with the description of PTSD found in the *Diagnostic and Statistical Manual of Mental Disorders* (DSM), published by the American Psychiatric Association and now in its fifth edition. Watters points out that the Sri Lankans do not have a Cartesian notion of mind-body separation and tend to express traumatic memories in terms of bodily pain. This is manifest in the tradition of Sinhalese *kolam* theatre and cleansing rituals such as the *Sani Yakuma*.[58] In them, the afflicted are encouraged to dance, shake, and shout unintelligible phrases, and the healers wear masks and dress as wild spirits and try to induce as much fear in them as possible. Watters reports that many people who complete these embodied cleansing rituals show dramatic recoveries.[59] In this respect, Aristotle's pronouncement in *Poetics* that *catharsis* is effected through the representation of *eleos* (pity/empathy) and *phobos* (fear) and other emotions (1449b27–28) is more closely related to the traditional performance therapies of *kolam* than to the clinical designations of the DSM. Even within cultures, attitudes toward psychological issues can change quite rapidly. In this respect, the development of the DSM stands as a marker of the way in which the prevailing culture creates and reinforces psychological disorders (e.g., homosexuality was listed as a mental disorder in the DSM until 1974).

We should also be cautious not to adopt the extreme relativist position either. While we accept that different cultures express psychological and emotional responses differently, we also see the same types of environmental stimuli producing similar biological and psychological responses. What we now call PTSD was called battle fatigue in World War II, shell shock in World War I, debility syndrome in the Boer War, and in the American Civil War, the physical chest pain that may have been caused by extreme anxiety was named "Soldier's Heart." We hear of distinct emotional responses by ancient Greeks to extreme situations throughout Greek literature, and we can still empathize with many of them. Surely, one cannot now read the chilling description of the abandonment of the Athenian wounded at Syracuse, as told by Thucydides (7.75), and not see an expert description of what we now call "survivor's guilt." Greek tragedy is replete with poignant moments that would have resonated directly with an audience who had been or were suffering the destruction and losses of war. Any veteran of combat, ancient or modern, reacts to these kinds of vivid descriptions of suffering and war from a place of deep personal experience. Consider the messenger from Aeschylus's *Agamemnon* who has just returned from 10 desperate years fighting at Troy:

You wouldn't believe what we had to put up with,
crammed into ships, jammed in narrow gangways,
berthed on bare boards, hardship every single day.
Then ashore it was even worse,
camped under the enemy walls, constantly
soaked through the skin, with damp and drizzle,
our clothes rotting, and our hair crawling with lice.
                Aeschylus, *Agamemnon*, 555–561

These words could have been uttered by a veteran of the America Civil War, from the trenches of the Somme, the freezing ground of the Ardennes in World War II, or the marshes of Southern Iraq.

Whereas David Kontsan has pointed out that we should not assume that Greek emotions map perfectly onto modern Western concepts, Fagan makes the essential point that we are able to recognize certain common behaviors in people from other cultures, including ancient ones. We can empathize with the emotionality of Greek tragedy, laugh with the clash of generations, genders, or political positions in much of Aristophanes, and relate to the dilemmas set down by Thucydides. Fagan writes: "The ability to identify with the behavior of people far removed from us chronologically, geographically, and culturally is fundamental to the business of history."[60] This can be equally applied to drama, though we may not completely understand every aspect of the Athenian emotional, scopic, and cultural regime for whom the work was created, we can still find ways in which to relate to the shock, pain, and suffering of tragedy or the humor of comedy. Something elemental is still being transmitted, even though we should also strive to interpret Greek drama through the prism of the culture it was intended for. In this case, cognitive theory provides some necessary distance and a degree of objectivity. This allows for the posing of fundamental questions, such as how does tragic narrative operate? What does it mean to perform in the open air? Why wear a mask, and how did it function? How did watching movement affect the audience? What happens when words become unintelligible? How did ancient drama create a sense of empathy?

Yet, while establishing a cultural distance is vital, this study also takes in the fact that in many ways Greek drama is the progenitor of the theatre we experience today. The form of Athenian theatre contains many of the same features we have come to associate with a theatrical performance: it used actors involved in live action, telling narratives within a certain temporal frame, set within mutually understood cultural, social, religious, and political formats. We can also observe how these forms were constantly being explored and expanded. This is why Greek drama provides a good model for a study that also seeks to emphasize the continuing importance of theatre in our society. Though we are different from the Greeks, we still share some of the same basic human traits that compel us to seek out theatricality, whether it may be found on our screens, in sporting events, rituals, courtrooms, political institutions, or theatres.[61] The key to caring about the theatre then and now is still the same: empathy.

Empathy

In the following chapters, I am going to make a case that Greek drama, and tragedy in particular, provided its audience with a multisensory experience that generated the ability for them to feel for characters in circumstances that went beyond their everyday experience. In order to do this, drama needed to provoke affective responses in its audience members and have them become cognitively absorbed in the action they were experiencing. Absorption, a state of focused immersion involving a narrowing of concentration and attention, is provoked by dissociation, which can in turn lead to greater feelings of empathy. Absorption often includes temporal dissociation (you lose track of time, or become totally absorbed in a fictitious temporal pattern presented on stage) and a sense of heightened pleasure. Self-concerns and the external world beyond the absorbing experience become irrelevant, and critical thought is curtailed.[62] Being "absorbed" by a piece of music, a play, or a film is reported by most people as being highly pleasurable, and yet one of the primary inducers of cognitive absorption is dissociation.

Dissociation is defined as the functional alteration of memory, perception, and identity, and it can occur in response to stress or trauma; within social rituals such as traditional healing practices and aesthetic performances; and in some extreme cases, spontaneously.[63] People have differing degrees of dissociation depending on a variety of social, cultural, and biological factors, but those who are more prone to dissociative experiences leading to cognitive absorption showed a higher capacity to enjoy, for example, negative emotions in music.[64] The most common form of dissociation is absorption, which Garrido and Schubert have described as "the heart of the normative dissociative process."[65] This seems like a contradiction in that something sad might actually give pleasure, but the feeling of absorption derived from this kind of cognitive dissonance can lead to people reporting pleasurable experiences. Through absorption, audience members (and likewise, readers of fiction) become more attuned to the predicaments and feelings of the fictional characters they are experiencing and, in short, they come to feel more empathetic.[66]

Plato was also concerned with this question in *Philebus* (48a), where he asks why people derived both pleasure and pain from songs of mourning, as was Aristotle, who noted how sad music produced a corresponding state of feeling in the audience (*Politics* 8.5–8). This notion of performance creating a "corresponding state of feeling" is also found in *Laws*, where Plato tells us how tragic playwrights, "with words and rhythms and music of the most morbid kind work up the emotions of the audience to a tremendous pitch, and the prize is awarded to the chorus which succeeds best in making the community burst into tears" (8.800d-e).

But what do we mean exactly by empathy? The word itself has become fraught and is frequently now applied as meaning "feeling with" in opposition to *sympathy* denoting "feeling for."[67] But the process of how the affective state of one person is perceived, evaluated, and "felt" by another must be more complex. First, can we really ever know the emotional state of another person? To be sure, we might be able to recognize the broad perceptual markers of say, sadness, but can we really fully comprehend how the sadness on display has been generated and is affecting the person we are viewing? When we say that we feel empathy for another, are we simulating their emotional state or are we actually being provoked by our cognitive processes to project our own experiential realities back onto the person under view? Empathy is commonly described as the ability to understand and share the feelings of another, but there is a big difference between understanding and sharing – the latter implies that we are able to feel the same as another person. Even "understanding" is a slippery term. Do we really understand another person's feeling or are we relating their experience with our own and forming a new affective response based on that evaluation?

Goldie tackled this empathy problem by positing that four elements needed to be in place for people to be able to share the same feelings.[68] First, there can be no difference in the psychological dispositions of each person. This is clearly impossible: so many factors go into forming a person's psychological disposition, including genetics and epigenetics, environment, family, life experiences, past traumas, upbringing, and education, to name but a few, that no two people can ever be said to have identical mentalities. Even siblings raised in the same environment can be very different in so many ways. Second, there can be no non-rational influences of the observed person's psychological make-up or decision-making process. This is also important because in order to share what they are feeling, we presumably need to understand the motivations and objectives that have led them to this place. While we might well share such non-rational elements, we can never know them all in another person. Third, there should be no significant confusions in the observed psychological make-up, and fourth is that there can be no psychological conflicts. The same provisos outlined in number two hold here as well. Goldie's point is that we can never know the interiority of somebody else's mind and the myriad factors that have led them to the very place where we are now observing them.

Slaby calls this failure of a basic empathy of shared emotions the "usurpation of agency."[69] He cites personal choice as an example of how one's own agency informs every decision someone makes. If we are sharing feelings with another person involved in a difficult choice, then we must be thinking through these choices from at least some perspective of what we might do in a similar situation. We therefore usurp the observed person's agency with our own, based on our own psychological make-up. Therefore, what we ultimately feel about the choices we observe being made by another may not resemble that person's feelings at all.

Slaby, Goldie and others propose that it is futile to try to reproduce in your own mind the affective mental state of another, and this poses a serious challenge to the basic definition of empathy as the sharing of feelings between different people. But this should not condemn the very idea of empathy, which I take to mean that one's own perspectives, affective states, or feelings can be altered by observing and evaluating the feelings of another. Where the problem lies is in the claim that interior mentalities can be simply shared as part of an affective perceptual process. The production and reception of theatre is an empathetic process if we apply the definition I use above. We have to care about the characters we observe and be interested in the choices they make; otherwise, the drama fails. The key here is we need to observe emotion in action – to be moved ourselves, we need to encounter the movement of other mentalities. Perhaps this idea is approximate to what Aristotle meant by *praxis* – the embodiment of drama in action.

Slaby places this idea of empathy in action under the phenomenological rubric of what is known as *interaction theory*, which proposes an alternative to simulation theories of empathy where one simply feels what the observed person is feeling. Instead of one interior mind interpreting or transferring another interior mind, interaction theory proposes mutual interactive engagement, an essential part of which is the ability to see affective states displayed facially, bodily, and aurally in plain view. He argues that these commonly understood affective markers are products of a joint agency and joint active world orientations and that "the only way to meaningfully engage with another person's mentality without imposition is by engaging with her on the level of action," what he calls "co-engagement," such as jointly striving toward a common goal or the completion of a joint project. Slaby cites de Jaegher's and di Paolo's concept of "participatory sense-making" and Krueger's "we-space" and calls them "realms of co-presence, of lived mutuality, bodily enacted between interacting individuals."[70] These theories of extended, distributed cognition map perfectly onto the act of theatre, which is such a "we-space" where actors and audience co-engage in active, participatory sense-making.

The theory of mutual interactive engagement is dependent on the observer having a basic understanding of the affective signals being displayed and communicated by the person being observed. Slaby points out that the mental states of another are not "hidden in the interior of another person, but instead something that is in plain view," and therefore experienced directly and without too much difficulty, hence the joint engagement nature of interaction theory.[71] This is analogous to Aristotle's concept of poetic universals in *Poetics*, wherein he describes a "universal" as "the sort of thing that a certain kind of person may well say or do in accordance with probability or necessity" (1451b 10–12). For Aristotle, this is what most distinguishes poetry from history, in that history deals with particulars and relates what has happened, whereas poetry tends to universalize what might happen. The key here is probability, in that the audience must be able to recognize the causal events and have a shared stake in the predictive nature of the representation of action (1451b 30–33).

From a neurological perspective, empathy has been described as a bilateral process involving autonomic somatic mirroring systems combined with an evaluative cognitive process.[72] In this appraisal, empathy is also not held as a binary to sympathy but is a more fluid process, whereby the affective states of the target are bodily mirrored, felt, and then evaluated based on contextual information. This empathy network has been described as "a vicariously sharing emotional system that supports our ability to share emotions and mental states and a cognitive system that involves cognitive understanding of the perspective of others."[73] This use of empathy also assumes that the emotions of the target have affected the other person, but that this person may not necessarily feel or display the exact same emotional state.

This hypothesis is supported by clinical practice in dance, music, and drama therapies, where therapists and participants use mirroring techniques to gain a better understanding of another's psychological situation, in order to inform and aid treatment. David Alan Harris has described a striking example of this empathetic process in practice in his account of his work with former child-soldiers in Sierra Leone.[74] Amongst many mirroring group exercises performed over time, each boy was asked to "direct" scenes of his former life using the other boys. One "little plot" to borrow Aristotle's term describing proto-tragedy from *Poetics* (1449a 20–21) involved the boys playing their own victims, such as mothers who had their babies killed in front of them and family members who were forced to watch relatives die. By mirroring and then embodying their most traumatic memories of what they were forced to do while pressed into combat at such a young age, they were able to begin to confront their own anger, guilt, and sadness, by performing the roles of those they had been forced to terrorize. At the end of the program, the boys volunteered to present such scenes, as well as ones showing their own traumatic recruitment into the child-army, to their community, which had hitherto justifiably feared the presence of these young former child-warriors. Harris reported that the performance in a festival environment caused a highly emotional response from the townspeople, who reported that the drama helped them to understand the trauma of those who had formerly traumatized them, and facilitated the re-establishment of a sense of trust and community inclusiveness. This is a practical example of the kind of bilateral (automatic/evaluative) empathetic response that in a similar way, and to a different extent, ancient tragedy was also capable of effecting. In Chapter Four I will examine this kind of cognitive mirroring as it related to watching movement, dance, and gestures and the role of kinesthetic empathy and emotional contagion in the experience of Greek drama.

Empathy then, as I am using the term here, is not the transference of an affective state from one person to another; rather, it is the ability to perceive emotional signals projected by another person and to apply our predictive cognitive mechanisms to seek to place those signals in some kind of comprehensible context. Drama relies on empathy working this way in order to involve its audience in the collective "we-space" that is the live theatre.

One famous example from Greek tragedy shows this process in action. In Sophocles' *Ajax*, Odysseus is turned into an invisible spectator by Athena, who allows him to experience the exultantly violent and deluded behavior of Ajax, who is torturing domestic animals thinking them to be his fellow Greek commanders, including Odysseus, who he believes robbed him of the arms of Achilles. In effect, Odysseus becomes the audience, Ajax the actor, and Athena the dramatist. When Odysseus is placed in this theatrical relationship with Ajax, he remarks that he feels scared – he does not at all share the crazed ecstasy of the man he is observing. He has, however, engaged in a form of participatory sense-making and concludes that he feels *eleos* ("pity" or "empathy") because of Ajax's cruel predicament, and says he is "not thinking of his fate, but my own" (125). At the end of the play, Odysseus applies the lesson he has learned as a *theates* (spectator) and tells a hostile Agamemnon that he should allow the burial of Ajax as he will also one day need to be buried. But this is also a political decision in that it allows discipline in the army to be maintained and Agamemnon to not feel his authority has been compromised. This transference of the observation of the predicament of another onto one's own personal experience makes theatre such a powerful social and political tool.

The mimetic mind

I hope that one of the results of this cognitive approach to ancient theatre is to start to view performance, and drama in particular, as a manifestation of how the culture it was created for thinks. Conversely, the theatre has often been used as a metaphor for human cognition. When Plato complains of the Athenian *theatrocracy*, he is equating the enfranchisement of political opinion with the popularity of the theatre; in *Poetics* Aristotle traces mimesis as one of the earliest forms of human interaction; and Cicero described the relationship of the theatre to the people of the late Roman republic as *teatro populoque Romano* – "the theater and the people of Rome."[75] More recently, Daniel Dennett has use the term "Cartesian Theater" to debunk the notion that the brain operates by processing sensory inputs and then reassembling them as some sort of projection in the mind. For Dennett this means that these representations would need to be observed by some sort of homunculus, a little man inside the brain who is able to interpret the information and then command the body to respond.[76] Dennett is ridiculing the idea of Descartes, who proposed that consciousness was the product of the immaterial soul interfacing with the brain's physical pineal gland, which Descartes observed was in the middle of the brain. Dennett points out that without accepting some form of the concept of an immaterial soul, much modern perceptual science that relies on the idea of external perception leading to internal representation would have to accept his notion of the Cartesian theatre. This in itself provides an infinite regress in that we would then ask what controls the mind of the homunculus and so on. Dennett's own views on how the mind processes information are not dissimilar to Clark's, in that a process of constant editorial revision creates a comprehensible narrative stream. Likewise, Clark and Eilan push the theatrical metaphor beyond Descartes and imagine the entire body, not just the brain, as the "Heideggerian Theater", the place where many of the "multiple, quasi-independent forms of internal (and external) representation and processing forms" come together.[77]

Dennett's Cartesian Theater is imagined as a cinema with the homunculus gazing at a screen, whereas Clark's and Eilan's Heideggerian analogy presents a distributed cognitive mechanism of brain, body, and environment. An imagined theatre to be sure, but one that operates in the same way as the actual experience of participating as an audience member in the act of a theatrical performance. In many ways, the experience of watching theatre and the process of making it are models of distributed cognition and predictive processing. Good drama compels us because it can operate within our existing cognitive systems and create new unexplored worlds, but theatre can only be successful with the complicity of an audience, who must be present in order to fully participate in the theatrical event. This book draws the conclusion that both the embodied presence of the theatre and the experience of being transported into unexpected, uncomfortable, and dissociative places was highly valued by the Athenians. For them, theatre combined cultic associations, civic identity, entertainment, and perhaps most importantly a mind-opening experience to alternative perspectives and ideas.

To conclude this introduction, I make an appeal: I believe that theatre's ability to transport its audiences across temporal, cultural, and social boundaries and to absorb them in the emotional worlds of other people is essential for a citizenry who are to be full participants in any functioning democracy and be able to have any understanding or empathy for people who are not like them. I also hope that by going back to the Greeks and looking closely at what made their theatre such a powerful cultural force, I am also tacitly making a strong case for embodied presence, not only in the performing arts, but also in education, political discourse, and interpersonal communication. At a time when more and more of our social and cultural interactions are being digitized and mediated by screens, perhaps the model of ancient Greek drama can act as an exemplar of the importance of human-to-human communal contact. Plato may have bemoaned the *theatrocracy* that he felt was empowering too many ordinary Athenians to dare to feel that they could express an equal opinion, but this is the very heart of a vibrant and properly functioning free society. We still need a theatre, in all its various forms, that can bring us together, challenge our expectations, transport us to other places, and most importantly allow us into the minds of others.

Notes

1　Plato *Laws* 3. 701–702; Aristotle *Politics* 3.1281b6–10.
2　Cartledge 1997: 19.
3　Source, The Broadway League. www.broadwayleague.com/index.php?url_identifier= calendar-year-stats-1. Accessed January 12th 2016.
4　Chaston 2010 and Mueller 2016. Also, to a certain extent, Revermann 2013.
5　Stanford 2014; Rehm 2002; Budelmann and Easterling 2010; Budelmann 2010; Cairns 2015 and 2016.
6　Habinek 2011.
7　Blair 2007; Tribble 2011; Rokotnitz 2011; McConachie and Hart 2006.
8　McConachie 2008; Lutterbie 2011; see also Shaughnessy 2013.
9　Tribble and Sutton 2011: 94–103. I thank Lambros Malafouris for sending me to this work.
10　Auslander 2008: 52.
11　Papastamati-von Moock 2015.
12　Appadurai 1988; Malafouris and Renfrew 2010.
13　Renfrew 2012: 130.
14　Clark 1997: 62.
15　Hutchins 1995.

16 Malafouris 2013: 67.
17 Malafouris 2013: 79.
18 Noë 2009.
19 LeDoux 1998: 29; Edelman 2006: 26.
20 Damasio 2006: 118; Damasio and Carvalho 2013: 143–152.
21 Nisbett 2003.
22 Anderson 2015: 46.
23 Clark 2013: 181–204; and 2015.
24 Clark 2015: xiv.
25 Clark 2015: xvi.
26 Plato *Minos* 231a; Isocrates *Evagoras* 2.49 and 2.10; Aristotle, *Poetics* 1450b.16–21.
27 Plato *Apology* 18b5–7.
28 Wiles 2007: 92–112.
29 Ford 2015.
30 Tarlow 2012: 170.
31 Lloyd 2007: 59–60.
32 Plato *Republic* 4.437d1–442c7; Aristotle, *Rhetoric* 2.1–11; Xenophon *Memorabilia* 2.6.21.2–23.7.
33 Stoklosa 2012: 72–93.
34 Prinz 2012: 242–247.
35 Chaniotis 2012: 14.
36 Cairns, Douglas 2008: 45–46.
37 Konstan 2006: 261.
38 Rosenwein 2010: 1–32.
39 Rosenwein 2010: 10.
40 On biocultural approaches to literature, see Boyd 2005: 1–23 and 2009.
41 See Elbert and Schauer 2014; Roth 2013; Whalley 2014.
42 Marcus 2012.
43 Previc et al. 2005
44 Habinek 2011: 216.
45 Stanford 1983.
46 Fagan 2011: 40–41.
47 Fagan 2011: 64.
48 See also Konstan 2006: 3–40.
49 Tomkins 1962. These are Joy, Excitement, Surprise, Anger, Disgust, reaction to a bad smell, Distress, Fear, and Shame. Ekman proposed six: Joy, Surprise, Anger, Fear, Disgust, Sadness. Recently, Rachel Jack has suggested four and posited that fear and surprise are congruent, as are anger and disgust (Jack 2013: 1248–1286).
50 Kaster 2005: 6–7.
51 Kaster 2005: 3.
52 Chaniotis 2012: 23.
53 Martin 2012: 168.
54 See Malafouris and Renfrew 2010: 1–12.
55 See Abramiuk 2012: 95–124.
56 Shay 1995 and 2003; Tritle 2002 and 2014: 87–104.
57 Crowley 2012 and 2014: 105–130.
58 Bailey 2006.
59 Watters: 110.
60 Chaniotis 2012:11–36.
61 Woodruff 2008.
62 Agarwal and Karahanna 2000: 665–694.
63 Seligman and Kirmayer 2008.
64 Garrido and Schubert 2011: 279–296.
65 Butler 2006: 45–62.
66 Wickramasekera 2007: 59–69; Eerola et al. 2016.

67 See, for example, Cairns 2015: 75–94.
68 Goldie 2011: 308.
69 Slaby 2014: 251.
70 De Jaegher and Di Paolo 2007:485–507; Krueger 2011: 643–657.
71 Slaby 2014: 255.
72 Decety and Michalska 2013.
73 Shamay-Tsoory 2014.
74 Harris 2007: 203–231.
75 Parker 163, "a theater which did not merely reflect the Roman social order but enacted it," Cicero, *Sestus* 106 and 116.
76 Dennett 1997.
77 Clark and Eilan 2006.

Bibliography

Abramiuk, M.A. 2012. *The foundations of cognitive archaeology*. Cambridge, MA: MIT Press.

Agarwal, R. and Karahanna, E. 2000. "Time flies when you're having fun: Cognitive absorption and beliefs about information technology usage" in *MIS Quarterly* 24.4: 665–694.

Anderson, M. 2015. *The renaissance extended mind*. New York: Palgrave Macmillan.

Appadurai, A. 1988. *The social life of things: Commodities in cultural perspective*. Cambridge: Cambridge University Press.

Auslander, P. 2008. *Liveness: Performance in a mediatized culture*. London: Routledge.

Bailey, M.S. and de Silva, H.J. 2006. "Sri Lankan sanni masks: An ancient classification of disease" in *BMJ* 333.7582: 1327–1328.

Blair, R. 2007. *The actor, image, and action: Acting and cognitive neuroscience*. London: Routledge.

Boyd, B. 2005. "Literature and evolution: A bio-cultural approach" in *Philosophy and Literature* 29.1: 1–23.

Boyd, B. 2009. *On the origin of stories*. Cambridge, MA: Harvard University Press.

Budelmann, F. 2010. "Bringing together nature and culture: On the uses and limits of cognitive science for the study of performance reception" in Hall, E. and Harrop, S. (Eds.), *Theorising performance: Greek drama, cultural history and critical practice*. London: Bloomsbury: 108–122.

Budelmann, F. and Easterling, P. 2010. "Reading minds in Greek tragedy" in *Greece and Rome (Second Series)* 57.2: 289–303.

Butler, L.D. 2006. "Normative dissociation" in *Psychiatric Clinics of North America* 29: 45–62.

Cairns, D. 2008. "Look both ways: Studying emotion in ancient Greek" in *Critical Quarterly* 50.4: 43–62.

Cairns, D. 2015. "The horror and the pity: Phrikē as a tragic emotion" in *Psychoanalytic Inquiry* 35.1: 75–94.

Cairns, D. 2016. "Mind, body, and metaphor in ancient Greek concepts of emotion" in *L'Atelier du Centre de recherches historiques: Revue électronique du CRH* 16. http://acrh.revues.org/7416; doi:10.4000/acrh.7416. Accessed November 11th 2016.

Cartledge, P. 1997. "Deep plays: Theatre as process in Greek civic life" in Easterling, P.E. (Ed.), *The Cambridge Companion to Greek Tragedy*. Cambridge: Cambridge University Press: 3–35.

Chaniotis, A. (Ed.). 2012. *Unveiling emotions: Sources and methods for the study of emotions in the Greek world*. Stuttgart, Germany: Franz Steiner.

Chaston, C. 2010. *Tragic props and cognitive function: Aspects of the function of images in thinking* (Vol. 317). Leiden, the Netherlands: Brill.

Clark, A. 1997. *Being there: Putting brain, body, and world together again.* Cambridge, MA: MIT Press.

Clark, A. 2013. "Whatever next? Predictive brains, situated agents, and the future of cognitive science" in *Behavioral and Brain Sciences* 36.3: 181–204.

Clark, A. 2015. *Surfing uncertainty: Prediction, action, and the embodied mind.* Oxford: Oxford University Press.

Clark, A. and Eilan, N. 2006. "Sensorimotor skills and perception" in *Proceedings of the Aristotelian Society, Supplementary Volumes*: 43–88.

Crowley, J. 2012. *The psychology of the Athenian Hoplite: The culture of combat in classical Athens.* Cambridge: Cambridge University Press.

Crowley, J. 2014. "Beyond the universal soldier: Combat trauma in classical antiquity" in Meineck, P. and Konstan, D. (Eds.), *Combat trauma and the ancient Greeks.* New York: Palgrave Macmillan: 105–130.

Damasio, A.R. 2006. *Descartes' error.* New York: Random House.

Damasio, A.R. and Carvalho, G.B. 2013. "The nature of feelings: Evolutionary and neuro-biological origins" *Nature Reviews Neuroscience* 14: 143–152.

Decety, J. and Michalska, K.J. 2013. "A neuroscience perspective on empathy and its development" in Rubenstein, J. and Rakic, P. (Eds.), *Neural circuit development and function in the healthy and diseased brain: Comprehensive developmental neuroscience* (Vol. 3). San Diego/London: Academic Press: 379–393.

De Jaegher, H. and Di Paolo, E. 2007. "Participatory sense-making" in *Phenomenology and the Cognitive Sciences* 6.4: 485–507.

Dennett, D.C. 1997. "The Cartesian theater and 'Filling In' the stream of consciousness" in Block, N.J., Flanagan, O.J. and Güzeldere, G. (Eds.), *The nature of consciousness: Philosophical debates.* Cambridge, MA: MIT Press: 83–88.

Edelman, G.M. 2006. "The embodiment of mind" in *Daedalus* 135.3: 23–32.

Eerola, T., Vuoskoski, J.K. and Kautiainen, H. 2016. "Being moved by unfamiliar sad music is associated with high empathy" *Frontiers in psychology* 7: 975–983.

Elbert, T. and Schauer, M. 2014. "Epigenetic, neural and cognitive memories of traumatic stress and violence" in Caprara, G. V and Alessandra, G. (Eds.), *Psychology serving humanity, proceedings of the 30th International Congress of Psychology: Volume 2: Western psychology.* London/New York: Psychology Press: 215–227.

Fagan, G.G. 2011. *The lure of the arena: Social psychology and the crowd at the Roman games.* Cambridge: Cambridge University Press.

Ford, A. 2015. "The purpose of Aristotle's poetics" in *Classical Philology* 110.1: 1–21.

Garrido, S. and Schubert, E. 2011. "Individual differences in the enjoyment of negative emotion in music: A literature review and experiment" in *Music Perception: An Interdisciplinary Journal* 28.3: 279–296.

Goldie, P. 2011. "Anti-empathy" in Coplan, A. and Goldie, P. (Eds.), *Empathy: Philosophical and psychological perspectives.* Oxford: Oxford University Press: 301–317.

Habinek, T. 2011. "Tentacular mind: Stoicism, neuroscience and the configurations of physical reality" in Stafford, B.M. (Ed.), *A field guide to a new meta-field: Bridging the humanities-neurosciences divide.* Chicago: University of Chicago Press: 64–83.

Harris, D.A. 2007. "Pathways to embodied empathy and reconciliation after atrocity: Former boy soldiers in a dance/movement therapy group in Sierra Leone" in *Intervention* 5.3: 203–231.

Hutchins, E. 1995. *Cognition in the wild.* Cambridge, MA: MIT Press.

Hutchins, E. 2006. "Imagining the cognitive life of things" presented at workshop, *The Cognitive life of things: Recasting the boundaries of the mind*. McDonald Institute for Archaeological Research, Cambridge. http://liris.cnrs.fr/enaction/docs/ documents2006/ ImaginingCogLifeThings.pdf.

Jack, R.E. 2013. "Culture and facial expressions of emotion" in *Visual Cognition* 21.9–10: 1248–1286.

Kaster, R.A. 2005. *Emotion, restraint, and community in ancient Rome*. Oxford/New York: Oxford University Press.

Konstan, D. 2006. *The emotions of the ancient Greeks: Studies in Aristotle and classical literature* (Vol. 5). Toronto: University of Toronto Press.

Krueger, J. 2011. "Extended cognition and the space of social interaction" in *Consciousness and Cognition* 20.3: 643–657.

LeDoux, J. 1998. *The emotional brain: The mysterious underpinnings of emotional life*. New York: Simon and Schuster.

Lloyd, G.E.R. 2007. *Cognitive variations: Reflections on the unity and diversity of the human mind*. Oxford: Oxford University Press.

Lutterbie, J. 2011. *Toward a general theory of acting: Cognitive science and performance*. New York: Palgrave Macmillan.

McConachie, B.A. 2008. *Engaging audiences: A cognitive approach to spectating in the theatre*. New York: Palgrave Macmillan.

McConachie, B.A. and Hart, F.E. (Eds.). 2006. *Performance and cognition: Theatre studies and the cognitive turn*. London: Routledge.

Malafouris, L. 2013. *How things shape the mind*. Cambridge, MA: MIT Press.

Malafouris, L. and Renfrew, C. (Eds.). 2010. *The cognitive life of things: Recasting the boundaries of the mind*. Cambridge: McDonald Institute for Archaeological Research.

Marcus, G.F. 2012. *Guitar zero: The new musician and the science of learning*. New York: Oneworld Publications.

Martin, L.H. 2012. "The future of the past: The history of religions and cognitive historiography" in *Religio* 20.2: 155–172.

Mueller, M. 2016. *Objects as actors: Props and the poetics of performance in Greek tragedy*. Chicago: University of Chicago Press.

Nisbett, R.E. 2003. *The geography of thought how Asians and westerners think differently – and why*. New York: Simon and Schuster.

Noë, A. 2009. *Out of our heads: Why you are not your brain, and other lessons from the biology of consciousness*. London: Palgrave Macmillan.

Papastamati-von Moock, C. 2014. "The theatre of Dionysus eleuthereus in Athens: New data and observations on its 'Lycurgan' phase" in Csapo, E., Goette, H.R., Green, J.R. and Wilson, P. (Eds.), *Greek theatre in the fourth century BC*. Berlin: Walter de Gruyter GmbH & Co KG: 15–76.

Papastamati-von Moock, C. 2015. "The wooden theatre of Dionysos eleuthereus in Athens: Old issues, new research" in Frederiksen, R., Gebhard, E.R. and Sokolicek, A. (Eds.), *The architecture of the ancient Greek theatre, acts of an international conference at the Danish Institute at Athens 27–30 January 2012*. Aarhus, Denmark: Aarhus University Press: 39–79.

Parker, H.N. 1999. "The observed of all observers: Spectacle, applause, and cultural poetics in the Roman theater audience" in *Studies in the History of Art* 56: 162–179.

Previc, F.H., Declerck, C. and de Brabander, B. 2005. "Why your 'head is in the clouds' during thinking: The relationship between cognition and upper space" in *Acta Psychologica* 1.118: 7–24.

Prinz, J.J. 2012. *Beyond human nature: How culture and experience shape our lives*. London: Penguin.

Rehm, R. 2002. *The play of space: Spatial transformation in Greek tragedy*. Princeton, NJ: Princeton University Press.

Renfrew, C. 2012. "Towards a cognitive archaeology: Material engagement and the early development of society" in Hodder, I. (Ed.), *Archaeological theory today*. Cambridge: Polity: 124–145.

Revermann, M. 2013. "Generalizing about props: Greek drama, comparator traditions, and the analysis of stage objects" in Harrison, G. and Liapis, V. (Eds.), *Performance in Greek and Roman theatre*. Leiden, the Netherlands: Brill: 77–88.

Rokotnitz, N. 2011. *Trusting performance: A cognitive approach to embodiment in drama*. New York: Palgrave Macmillan.

Rosenwein, B.H. 2010. "Problems and methods in the history of emotions" in *Passions in Context* 1.1: 1–32.

Roth, T.L. 2013. "How traumatic experiences leave their signature on the Genome: An overview of epigenetic pathways in PTSD" in *Frontiers in Psychiatry* 5: 93–93.

Salvo, I., Cairns, D.L. and Fulkerson, L. (Eds.). 2015. *Emotions between Greece and Rome* (BICS supplement 125). London: Institute of Classical Studies, School of Advanced Study, University of London.

Seligman, R. and Kirmayer, L.J. 2008. "Dissociative experience and cultural neuroscience: Narrative, metaphor and mechanism" in *Culture, Medicine and Psychiatry* 32.1: 31–64.

Shamay-Tsoory, S.G. 2014. "Dynamic functional integration of distinct neural empathy systems" in *Social Cognitive and Affective Neuroscience* 9.1: 1–2.

Shaughnessy, N. (Ed.). 2013. *Affective performance and cognitive science: Body, brain and being*. London: Bloomsbury.

Shay, J. 1995. *Achilles in Vietnam: Combat trauma and the undoing of character*. New York: Simon and Schuster.

Shay, J. 2003. *Odysseus in America: Combat trauma and the trials of homecoming*. New York: Simon and Schuster.

Slaby, J. 2014. "Empathy's blind spot" in *Medicine, Health Care and Philosophy* 17.2: 249–258.

Stanford, W.B. 2014. *Greek tragedy and the emotions (Routledge Revivals): An introductory study*. London: Routledge.

Stanford, W.B. 1983. *Greek tragedy and the emotions: An introductory study*. London: Routledge.

Stoklosa, A. 2012. "Chasing the Bear: William James on sensations, emotions and instincts" in *William James Studies* 9: 72–93.

Tarlow, S. 2012. "The archaeology of emotion and affect" in *Annual Review of Anthropology* 41: 169–185.

Tomkins, S.S. 1962. *Affect, imagery, consciousness: Volume I: The positive affects*. New York: Springer publishing company.

Tribble, E.B. 2011. *Cognition in the globe: Attention and memory in Shakespeare's theatre*. New York: Palgrave Macmillan.

Tribble, E.B. and Sutton, J. 2011. "Cognitive ecology as a framework for Shakespearean studies" in *Shakespeare Studies* 39: 94–103.

Tritle, L.A. 2002. *From Melos to My Lai: A study in violence, culture and social survival*. London: Routledge.

Tritle, L.A. 2014. "Ravished minds in the ancient world," in Meineck, P. and Konstan, D. (Eds.), *Combat Trauma and the Ancient Greeks*. London: Palgrave Macmillan: 87–104.

Whalley, K. 2014. "Epigenetics: Early trauma alters sperm RNA" in *Nature Reviews Neuroscience* 15.6: 349–349.

Wickramasekera, I.E. 2007. "Empathic features of absorption and incongruence" in *American Journal of Clinical Hypnosis* 50.1: 59–69.

Wiles, D. 2007. "Aristotle's poetics and ancient dramatic theory" in McDonald, M. and Walton, M. (Eds.), *The Cambridge companion to Greek and Roman theatre*. Cambridge: Cambridge University Press: 92–112.

Woodruff, P. 2008. *The necessity of theater: The art of watching and being watched.* Oxford: Oxford University Press.

1 *Muthos*

Probability and prediction

Aristotle famously describes *muthos* (plot/narrative) as "the soul of tragedy" (*Poetics* 1450a38) and narrative in action (*praxis*) as the aim of the art form (1450a23–26). For Aristotle, narrative should be well ordered and structured: in nature, as in drama, everything that is beautiful (*kalos*) must be perceptible. For example, if an animal is too small, we cannot see all of its constituent parts, just as we would be unable to perceive the totality of a 1,000-mile-long creature. In both cases we would be unable to discern their beauty. As the parts of organic structures need to be perceptible, so dramatic narratives ought to be well structured and not too long as to strain human memory (*Poetics* 1450b35–1451a6). But a unified structure is not enough on its own; the narrative needs to grab our attention, and it does this by means of challenging our expectations and arousing our sense of *thaumasios* (amazement). This in turn can lead to tragedy's most powerful attribute and the one that made Plato rail against it – its ability to arouse the extreme emotions of its audience. Aristotle lights on *eleos* (empathy/pity) and *phobos* (fear) and remarks that these and other emotions are most highly provoked when the incidents of the narrative are unexpected by way of such devices as reversals (*peripeteia*) or recognitions (*anagnōrisis*). Yet Aristotle is not advocating for a theatre of complete surprisal; he writes that the most amazing are not those that seem completely random, but ones that "appear to have happened incidents as if it were on purpose." His example is of the statue of Mitys at Argos that fell on and killed the murderer of the real Mitys. Therefore, actions that divert from the play's structure momentarily and cause amazement, but then make sense within the larger world of the narrative, are the most effective (*Poetics* 1452a2–11). In many ways, what Aristotle is describing is probabilistic thinking and prediction.

In this chapter I re-examine this famous aspect of *Poetics* and suggest two main ideas: (1) that Aristotle is articulating a cognitive feature of Athenian society, one that is key to understanding the importance of its theatre – the concept of *eikos*, or probabalistic thinking; and (2) that Aristotle's views on how a dramatist might create effective emotional theatre comport with contemporary theories of predictive processing. This is a unified theory of the mind that blends Bayesian and computational views of human cognition with philosophies of distributed cognition, including embodiment and enactivism, and it is the theory that underpins this cognitive study of ancient Greek drama. In particular, I want to focus on tragedy's ability to evoke a sense of wonder (*thaumasios*) by engaging probabilistic thinking (*eikos*), which can elicit powerful empathetic emotions (such as *eleos* and *phobos*), which in turn can result in a mind-changing experience (*catharsis*). Greek drama sought to offer its audiences alternative viewpoints and different perspectives in an attempt to change entrenched mindsets. As we shall see, it did this via dissociation, attention, affect, and empathy in a multisensory experience involving environment, masks, movement, music, words, costumes, and objects. In order for these elements to be effective, they needed to be contained within a cognitive scaffolding that corresponded to the mentality of the audience the work was intended for. This was *muthos* – a story replete with revelations, reversals, and wonderment.

One of the means by which Greek drama provoked probabilistic thinking was *peripeteia*, a reversal or sudden change. Stephen Halliwell has described Aristotle's concept of *peripeteia* in drama as being to "surprise us and arouse our minds to look for an underlying explanation of the ostensibly inexplicable."[1] Halliwell identifies the attainment of a sense of wonderment as an underlying concept that unites Aristotle's arguments in *Poetics*. He reminds us that Plato has Socrates declare in *Theaetetus* (155d) that "wonder is the source of philosophy," and that Aristotle writes in *Metaphysics* "it is through wonder that people originally began to philosophize" (982b11–12).[2] I want to unpack what exactly Aristotle was getting at here and suggest that he identified how a well-crafted drama exploited the human cognitive perceptual systems that are provoked by novelty and prediction.

Making sense of surprisal

Predictive processing is a theory that has been developed through recent work in cognitive theory and computational neuroscience, which posits that the human brain is an active prediction generator constantly processing the bottom-up signals transmitted by our somatic sensory systems, which are actively compared to top-down models. Whenever a sense signal conflicts with the stored perceptual model, an "error correction" occurs involving attention, action, and inference. This theory developed to explain how the brain is able to quickly perceive the world without using the massive amount of energy requited to analyze and reconstruct every bit of sensory data that comes flooding in at every given moment of our existence (even in sleep). This is a world described by Andy Clark as made of patterns of expectation, "in which unexpected absences are as perceptually salient as any concrete event, and in which all our mental states are coloured by delicate estimations of our own uncertainty."[3]

In his study of the communicative elements of audio-visual media including the theatre, Marshall Poe describes the human mind as "a remarkable pattern-building and pattern-recognizing machine." The kind of perceptual structural disruptions that Aristotle describes, Poe calls intrusive signals and categorizes them as anomalies and puzzles. Hence, nothing will draw human attention like something that should not be there – "your ears and eyes, then, are designed to draw your attention to anomalies and make you investigate them whether you like it or not."[4] Poe

echoes Aristotle's comments on the human predilection to seek perceptual structure but also to be amazed by disruptions to those structural forms. In the *Poetics*,
Aristotle wrote that theatrical works are called dramas because they represent
humans doing (*mimountai drõntas*) and that *mimesis* (representation) is natural to
human beings. Humans learn as they observe and infer what each thing is, and that
tragedy, comedy, and epic poetry were generated out of this distinctively human
natural trait (1450a15–17). Aristotle's ideas on *mimesis* are not at all dissimilar to
Andy Clark's description of how the human mind uses predictive processing to
sense the world and infer its meanings.

Predictive processing is connected to Friston's free energy principle, which
states that any self-organizing system that is at equilibrium with its environment
must minimize its free energy to resist a natural tendency to disorder or entropy.[5]
Maintaining low entropy means, in effect, minimizing states of surprise. Friston's
example is a fish out of water, which would be in a surprising state biologically
and therefore in peril of great entropy, leading to its demise as a functional biological system. Therefore, the free energy principle has it that all living organisms
must minimize the probability of encountering surprising states. What constitutes
"surprisal" is based on each organism's particular biological expectations, which
are generated by its own internal model of its world. Friston's fish has adapted to
a world of water, and the entropy generated by the massive surprisal of the total
removal of that world can prove fatal. Except that the fish's own somatosensory
systems autonomically communicate the surprisal and the fish's nervous system
produces the physical reaction of extreme bodily movements. These provide its
only hope of convulsing back into the water. Thus, the extreme action generated by
extreme entropy might mean that Friston's fish out of water sometimes gets away.

Friston has suggested that the free energy principle must be highly hierarchical
to be able to react to the multi-layer structure of the causal dependencies of its
world. For the fish, this means it will react bodily more violently the longer it is
out of water until complete entropy occurs. The fish has "learned" this behavior
through its and its evolutionary ancestors' interaction with its environment, which
means that while its mental predictions of how to react in each surprising situation are stored "a priori," the predictive model itself is "the sedimentation of 'a
posteriori' information." Thus, predictive processing is the interplay of a system
of distributed cognition, or as Clark puts it, "since the causal web of interactions
experienced by the [fish] has come effectively to be mirrored by, or embodied in,
the [Fish's] structure/function, it is appropriate to say that the [Fish] is, rather than
has, a model of its world."[6] If, like Friston's fish, we are also seeking to minimize
free energy and avoid surprise, then what do we make of Aristotle's views that it
is the moment of profound surprisal – amazements – that produce the empathy
and fear we need to feel to experience the *catharsis* of tragedy? This is exactly
how Friston's fish "learned" to flop violently about in a last-ditch and sometimes
successful attempt to get back in the water – the fish is desperately trying to correct the error of entropy.

What Aristotle names *thaumasios*, Poe called anomalies, and Friston surprisal,
Clark terms "error correction." We need to understand this process to get a better

grip on how theatre works and what I think Aristotle was attempting to describe. I have already explained how the living organism must minimize free energy to avoid entropy, but how does this work in practice? Clark provides a neat analogy with computer data compression, which minimizes bandwidth by predicting the pattern of the image it is compressing and sending. Rather than sending every single pixel that makes up the original image, it looks for the differences between pixels by finding the boundaries and then predicting where the pixels of the same color will be. As Clark explains, "it is the deviations from what is predicted that then carry the 'news', quantified as the difference (the "prediction error") between the actual current signal and the predicted one." This is the same process our digital TVs and computers utilize when we stream a film: almost all of the data required to reconstruct the image was already present in the previous frame.[7]

Predictive processing works in the same kind of way: we conserve free energy by creating models of the world around us based on our prior sensory experiences, and we then constantly modify these models as new sensory data is interpreted. We too cannot possibly process every detail of the world at all times, because we would quickly reach a state of sensorial overload and free energy entropy. Neuroscientist Chris Frith contends that what we really perceive are the brain's models of the world or "fantasies that coincide with reality."[8] In effect, the brain is making a best guess about the bottom-up sensory information it receives. If the signal is not very "noisy" in cognitive terms – it confirms quickly to the predictive model – then it passes easily up the predictive error correction chain to form a percept. If, however, the signal is noisy and confounds the predictive model, then the system demands that we pay attention to the "error" or the surprise and correct it before it can be passed on up to form a percept. Clark describes this process as a multi-level bidirectional cascade of top-down probabilistic generative models.[9] Aristotle's description of the power of *thaumasios* is crafted along similar lines: the best plot surprises are ones we are able to resolve into some sort of predictive pattern "since the most amazing [incidents] even among random events are those which we perceive to have happened as if it were on purpose" (1452a6–8).

Aristotle is advising playwrights to deliberately sprinkle their plots with surprising, if ultimately predictable, twists and turns, but doesn't this clash with Friston's free energy principle, one of the methodological underpinnings of predictive processing? This problem has been called the "darkened room puzzle" in that if surprisal causes such a large expenditure of cognitive energy, why do we not seek out an unsurprising as possible sensorial environment, namely a darkened room, and just stay there? This sounds quite similar to Plato's famous allegory of the cave in the *Republic*, where the people are imprisoned in a dark cavern and watch images of what they think is reality but is really just shadows flickering on the cave wall (514a-520a). It is the actions of the one cave dweller who breaks his bonds and ascends into the light to seek the truth that have the power to alter the perception of those left in the dark – if, that is, Plato's *theoric* explorer is brave enough to return to tell them what he has experienced. This human instinct to escape from the darkened room has been described as "novelty seeking, exploration, and, furthermore, higher level aspirations such as art, music, poetry or humor."[10]

Jakob Hohwy points out that locking yourself in a dark, silent room will only produce "transitory free-energy minimization, as the demands of the world and the body will not be avoided for long."[11] In other words, we need to be able to handle surprisal as a necessary function of basic survival for finding food, shelter, companionship, and protecting ourselves from external and sudden threats. We tend to prefer a cognitive middle ground somewhere between the fully predictive and unchanging dark room and the often dangerous and randomized world in which we need to live to survive. For example, infants focused on events neither too easy to predict nor too difficult preferred a moderate amount of information processing while avoiding wasting cognitive resources on events that were too simple or too complicated.[12] This is also reflective of Aristotle's comments that tragic poets first staged stories "at random" but that then they tended to cluster their stories around a few famous mythological households to create the best art out of this kind of narrative structure (1453a17–23). For Aristotle, Euripides exemplified this kind of poetic structuring, and he described him as "the most tragic of poets" (1453a30).

This notion is not unique to Greek theatre; similar ideas can be found across a variety of performance cultures and periods. For example, in *The Empty Space*, director Peter Brook warned that in the theatre we can try to "capture the invisible but we must not lose touch with common-sense – if our language is too special we will lose part of the spectator's belief."[13] If the performance is calibrated just right, Brook writes, then it can create "the happening effect," which he describes as "the moment when the illogical breaks through our everyday understanding to make us open our eyes more widely." Brook continues:

> The moment of surprise is a jolt to the kaleidoscope, and what we see in the playhouse we can retain and relate to the play's questions when they recur transposed, diluted and disguised, in life.[14]

Aristotle also advocated surprisal for theatre to function well, but, like Brook, in the right predictive amount. Surprisal demands cognitive attention, but if the novelty cannot be rationalized by cognitive error correction, then our minds may lose their grip of the predictive process, and we then lose interest in the narrative action of the play we are watching. We also find very similar ideas in the writings of Zeami, a 14th/15th-century practitioner of Japanese Noh drama. In his first treatise on drama, *Fushikaden*, Zeami explained the performance concept of *Hana* or the flower. This was the process of combining the forces inherent in the surrounding environment with the performer's own internal nature to present a completely new aspect appropriate to each moment as proscribed by the play.[15] If the flower is achieved, then the audience will be enthralled. Zeami linked this performance concept with *omoskiroki* or fascination, stating that it is the performer's task to amaze the audience members. The flower also extends to the concept of *mezurashiki*, or novelty, in that it should be fresh, new, and exciting. Noh actors tried to re-create the pure joy a person feels when contemplating a flower by their interactions with each other and the spectators creating a sense of novel excitement. Thus, according to Zeami, there were two basic conditions to achieving the flower: (1) the enthrallment of the audience via fascination – *omoskiroki*, and (2) the means to surprise the audience with novelty – *mezurashiki*. Zeami wrote:

> Although this art entails handing on the effects of tradition, something about it stems from individual capacity, which is therefore beyond words. I call it transmitting the flower through effects and attitudes, because it is a matter of the flower being transmitted from mind to mind through the attainment of its effects – plan to evoke unexpected excitement in people's minds – this is the flower.[16]

Aristotle's *thaumasios* as applied to plot construction sounds very similar to Zeami's *Hana* as applied to acting. Noh drama certainly shares many performative commonalities with Greek theatre, most important that they were both masked and relied on the expert performance of movement and gesture to convey emotionality; we will return to these elements later and profit by comparing them. In detailing the creative process in making drama, both Aristotle and Zeami seem to be articulating a basic premise of human perception based on surprisal – the way in which this error correction drives attention, the fundamental concept in the theory of predictive processing. In the theatre, the performer must keep our attention, and surprisal demands attention and cognitive resolution – we need to make a predictive inference on what we sense in the world, whether that world is real or a mimetic theatrical creation.

The importance of surprise on our perceptual systems was also widely discussed by phenomenologists such as Husserl, Heidegger, and Merleau-Ponty. Husserl's notion of a binary perceptive notion between "fulfillment" (*erfüllung*) and frustration (*enttäuschung*) shares much in common with Zeami's concept of the Flower and Aristotle's ideas on plot construction, in that while surprisal is essential for cognition, unresolved surprise is highly frustrating. For Husserl, the experience of fulfillment occurs when a subsequent perception agrees with a prior expectation, what he termed "togetherness" (*zusammengehörigkeit*) – a state of consciousness between what we sense and what we had anticipated.[17] Likewise, Françoise Dastur maintained that "phenomenology privileges neither the interiority of expectation nor the exteriority of surprise . . . it tries to think the strange coincidence of both."[18] She placed Heidegger's term *ereignis* in this context as meaning both "happening" and "appearing to view." Her description of the importance of surprise ("the event") in phenomenology is well worth reproducing here.

> The event in the strong sense of the word is therefore always a surprise, something which takes possession of us in an unforeseen manner, without warning, and which brings us towards an unanticipated future. . . . The exteriority of the event introduces a split between past and future and so allows the appearance of different parts of time as dislocated. The event produces, in the literal meaning of the word, the difference of past and future and exhibits this difference through its sudden happening. The event constitutes the "dehiscence" of time, its coming out of itself in different directions, which Heidegger calls "ekstasis."

This non-coincidence with oneself which allows the possibility of being open to new events, of being transformed by them or even destroyed by them, is also that which makes of the subject a temporal being, an existent being, a being which is able constantly to get out of itself. Openness to the accident is therefore constitutive of the existence of the human being.[19]

Husserl and Heidegger also placed surprisal at the core of human perceptual experience. Husserl's "frustration" is Heidegger's *ekstasis* – the very core of what it means to be human, the notion of being "open to accident." As Dastur argues, "the paradoxical capacity of expecting surprise is always in question in phenomenology."[20] The Greek sense of the word *ekstasis* means to be "out of body" or "beside yourself," and I will try to show here that one of the primary aims of Greek drama was to create its own mimetic "dehiscence of time" (something that puts the flow of time out of joint). This fostered a sense of dissociation, which led to enhanced empathetic responses that opened audiences up to new and alternate ideas and possibilities.

Merleau-Ponty proposed the same kind of perceptual reciprocal bottom-up/top-down system in phenomenological terms, claiming that we experience a percept in action (*pratiquement*) rather than by knowing it beforehand. He used the example of looking at a table lamp. In order to perceive the back of the lamp, which is not in our immediate visual field, we have to make an assumption based on the sensory information we do have that the rest of an object will confirm to our prior knowledge of what should be there. We can see the front of a lamp and assume we know what the back of the lamp looks like based on this kind of perceptual anticipation or prediction. Although it is a truism that the lamp has a back, "it is true" does not correspond to what is offered by perception: "perception does not give truths . . . but presences" and is a "synthesis of transition."[21] Merleau-Ponty elaborated on this concept of perception by explaining it in terms of the body in the world: in perceptual terms, our bodies have a double function – one is its various sensory fields, the other its entire body schema, which is "predestined to model itself on the natural aspects of the world."[22] This notion of sensory modeling is akin to Clark's predictive "bootstrap heaven," where the world provides the training signal – a form of self-supervised learning "in which the 'correct' response is repeatedly provided, in a kind of ongoing rolling fashion, by the environment itself."[23]

To emphasize this point in theatrical terms, we return once again to Japanese Noh drama and a dialogue between the Zazen priest Fausto Taiten Guareschi and the neuroscientist Giuseppe Pagnoni:

> Phenomenologically, the situation can perhaps be likened to the rarified intensity of Japanese Nō theater, where a prolonged stillness makes the appearance of a subtle gesture extremely salient. By maintaining the posture, we seem to set up a gap between expectation and its confirmation, an act that not only makes the cycle of expectation and confirmation-seeking itself more transparent to the meditator, but that also attenuates with time its reductive tension

(the tension to reduce the gap). More specifically, the process of active inference, which strives to select a winning hypothesis by literally picking and choosing from the sensorium, appears abated during meditation. This could correspond to a flattening of the probability distribution over many priors encoding for different hypotheses, which may in turn facilitate mental flexibility and creativity by weakening the dominance of long held beliefs in favor of less frequently considered alternatives.[24]

Aristotle takes a more nuts-and-bolts approach, but he is essentially advocating the same thing, that "the poet should put what is amazing (*thaumasios*) into his tragedies but they should not be *apithanos* or too incredible" (1460a13–30). In order to elicit empathy or pity (*eleos*) and fear (*phobos*), Aristotle lays out several ways to achieve novelty in the plot, even when the basic story is a traditional one and already known to the audience. Either they should use "invented" stories or use the traditional stories in a novel way (1453b13–37). His best example is when a character is about to do something terrible in ignorance of a relationship but recognizes it just before acting. He cites a tragedy by Euripides, *Cresphontes* (now lost): Merope, the Queen of Messenia, is just about to bring an ax down on the person whom she thinks killed her son. The audience knows that this is in fact her son. Plutarch tells us that this plot caused mass panic in the theatre and that the spectators jumped to their feet in terror.[25] For Aristotle, what facilitates *thaumasios* but not *apithanos* (unbelievable) is *eikos* or "probability," a concept he describes in *Rhetoric* as first being introduced as a device in persuasive speech by Corax of Syracuse (2.24.11). *Eikos* is also the term Aristotle uses in his first description of *Peripeteia* in *Poetics* (1452a24) to describe dramatic probability, it was an essential concept in Athenian thought and even the development of their democratic institutions, including the theatre. This is important in understanding the cognitive regime of Athenian drama, for as Victoria Wohl has stated, "tragedy is the genre par excellence of *eikos*."[26] Indeed it was – a mimetic mind that reflected the cognitive regime of the audience it was intended for, one for whom deeply embedded in their culture was probabilistic/predictive thought.

Wohl's assertion that *eikos* was an important part of the narrative structuring of Greek tragedy is certainly reflected by Aristotle, who in *Poetics* is interested in exploring the cognitive, even biological nature of different genres of *mimesis* rather than attempting to create some sort of manual on how to write plays for the budding dramatist. *Poetics* has been derided by theatre scholars in particular based on this latter idea, but it seems that Aristotle is really interested in the way in which mimetic forms elicit emotions that have real persuasive force, and *eikos* and *thaumasios* are some of the major players on this cognitive stage. This is why Aristotle writes "it is obvious from what we have said that it is the function of a poet to relate not things that have happened, but things that may happen, i.e. that are possible in accordance with probability (*eikos*) or necessity" (*Poetics* 1451a36-b1). The use of *eikos* in the tragedies of Sophocles has been studied by John Kirby, who comments that *eikos* carries with it "an implicit inference that allows the listener to make the logical connection him/herself based on what seems likely" and "serves an important psychological function in all human communication."[27] Also, Wohl has closely examined Euripides' *Helen* in terms of dramatic *eikos* and noted that the play, which centers around the image (*eikon*) of Helen in Egypt, "presents a sophisticated meditation on the force of probability within fiction and life."[28]

There is a good example of this concept of *eikos* in action in Aeschylus's *Agamemnon* (575). A messenger from the Argive army has come home after 10 years away with news of the return of Agamemnon. Yet his news is bittersweet, and he tells the chorus "some things turned out well, others went badly" (553). He recounts the hardships of war, the dead, the terrible conditions at Troy, and he then weighs this against the Argive victory and announces "it is appropriate (*eikos*) to boast under the sun" because when other men hear this "men will respect this city, praise our commanders, and honor Zeus" (575–580). The herald's boast is an example of the active and distributed nature of *eikos* in Greek society described by Kraus as making a claim acceptable to an audience "by pointing out a certain coherence and congruence of the speaker's own narrative with the audience's pre-established set of convictions, i.e. their ordinary everyday experience, their moral values, intellectual knowledge, emotional predispositions and behavioural habits."[29] This is reflected in a statement found in the *Rhetoric to Alexander*, attributed to Anaximenes of Lampsacus (1428a25): "a probability (*eikos*) is a statement supported by examples present in the minds of the audience." Thus, mimetic or rhetorical *eikos* is essentially a distributed cognitive feedback loop, constantly testing bottom-up new probabilities/possibilities against top-down predisposed convictions based on the audience's cultural views, emotions, and natural impulses. These three "types" of *eikota* are listed by Anaximenes (1428b5–10) as *pathos* (emotion), *ethos* (cultural habit), and *kerdos* (impulse).

The concept of *eikos* and its relationship to probabilistic thinking has been much discussed by scholars. In his influential work, *The Emergence of Probability*, Ian Hacking traces the development of probabilistic thinking to the 17th century, leading many scholars to question whether *eikos* should be associated with such a concept.[30] For example, David Hoffmann argues that the ancient Greek concept of *eikos* must have differed from what he calls our own frequency-based understanding of probability and suggests instead that *eikos* means "to be similar" or "resemble."[31] Manfred Kraus also rejects probability as a viable term because the mathematically calculable quantities would have been "completely alien to the ancient Greeks" and prefers to translate *eikos* as "plausibility."[32] But probabilistic thinking is not just the preserve of statistical inquiry; it can also be an important factor in mental conceptualization. In fact, it is one of the basic underpinnings of the theory of predictive processing, which in turn owes a great deal to what was originally a form of statistical inquiry – Bayes' theorem. Yet, while Bayes' theorem was originally devised to explain probability in games of chance, predictive processing is built upon the idea that in many ways the brain is Bayesian. If we place the Greek concept of *eikos* within a Bayesian cognitive framework, we can see that probability can mean a great deal more than just the statistical frequency of numbers.

Basic Bayes

In order to place *eikos* in its cognitive context, we need to take a brief tour of Bayes' theorem. I want to state at the outset that I am not proposing that the human brain is Bayesian (what that means exactly will hopefully become clear), but rather that the mind uses Bayesian type probabilities in accomplishing the rapid bottom-up/top-down/error correction cascade of predictive cognition. The concept of *eikos* is linked by Aristotle to *peripeteia* and amazement/surprise (*thaumasios*), and I intend to place this within the cognitive context of how humans, both ancient and modern, evolved to perceive and process the world around them.

Predictive processing suggests that we constantly make best guesses about the sensory information the world presents to us. We also put our highly valuable cognitive energy to work whenever we encounter an anomaly or surprise. In actuality, our very survival depends on being able to successfully manage and process these moments of cognitive novelty, and we are predisposed to seek out situational vicarious thrills in child's play and in the way we maintain interest in gripping narratives, extreme affective representations, and intriguing visual stimuli. That is, just not too much: for when novelty becomes unexplainable, our predictive systems fail and we either do not comprehend what we are perceiving or we make a prediction that is actually incorrect. A famous example of this is Gregory's hollow mask experiment, which we will discuss in greater detail in Chapter Three on masks. When a mask is revolved, the concave hollowed-out back of the mask suddenly transforms into a convex three-dimensional face – the front of the mask. You can watch the experiment for yourself on one of the websites listed in this note.[33] Such a simple, visual moment is truly bewildering. If you watch it again, even when you know that your predictive system failed on the first viewing, you are still unable to correct it the second time around, and the hollow mask still morphs from concave to convex. Clark proposes that this and other optical illusions are "optimal percepts" – the price we pay for "being able to get things right most of the time, in a world cloaked by ambiguity and noise."[34] But if we know it's an error, why can't we seem to correct that error on subsequent viewings? Why does our predictive architecture continue to fail us? I think there are some perceptive moments where our cognitive systems have evolved to sometimes defeat our predictive process, and we are unable to "bootstrap" our way to comprehension. Our perceptive attitude to faces is one such survival mechanism, which I think can help explain the profound emotional impact of Greek dramatic masks and greatly contribute to our understanding of the affective power of ancient drama. But much more will be discussed about the uncanny attributes of masks in Chapter Three.

What is revolutionary about predictive processing is that it reverses the traditional idea of perception, which views the brain as a passive receptor of sensory information motivated by bottom-up stimuli that cause cognitive responses. Instead, the generative model of predictive processing posits that our neural systems are actively trying to match the sensory information by sending top-down predictions to meet the bottom-up perceptual signals. This is instead of traditional top-down perception where all sensory information encountered is processed and responded to. With predictive processing only sensed deviations from what the brain has predicted need to be processed, and this saves an enormous amount of cognitive energy. Under the Bayes' rule of probability, our cognitive models, which have been formed by prior beliefs based on sensorial contact, are able to be constantly modified and corrected as new sensory information is processed. Basically, what a Bayesian system does is allow our pre-conceived ideas to be modified or changed in the face of conflicting information.

To get a grip on Bayes' theorem, we need to take a very brief detour into the world of statistical probability. Bayes' rule, as it is widely known, is named after Rev. Thomas Bayes (1701–1761), an English statistician, philosopher, and clergyman who developed a theory, published by Richard Price in 1763, that used probability rather than frequency to calculate outcomes. At its most basic, Bayes' rule allows new evidence to update prior beliefs. Versions of Bayes' rule were used by Alan Turing to crack the Enigma code in World War II, to identify the author of 11 disputed Federalist Papers as James Madison, and as statistical evidence in a number of prominent legal cases.[35] The simplest example I can cite is known as the Monty Hall problem, named after the TV host of NBC's *Let's Make a Deal* in the 1970s and 80s. Our host shows you three closed doors: behind one is the prize of a brand-new car, and behind the other two are goats. If you select the door that hides the car, then you get to drive it home, but if not you get a goat. Monty asks you to choose a door, and you select the middle door. He then opens the door on the left to reveal a goat because he knows what is behind each door. There are now only two closed doors, the middle door you selected and the one to the right. Now Monty asks you if you want to change your mind and switch your choice of doors. What should you do? Your rational self might decide that it makes no difference whether you stay with your first choice or change to the right-hand door. Is this true? Let's run the basic Bayesian numbers and find out.

When you made your choice, before Monty opened one door, you had a one in three chance of winning the car. Even though Monty opened a door, these odds will not change for you because one door has been opened. But wait, there are only two doors left, so you must now have a one in two or 50/50 chance of winning the car, right? Wrong. You have done nothing to impact your circumstances since making your initial choice, and the fact that Monty opened a door does not change your initial odds. You have not altered anything, and your odds stay the same – one in three or 33%. But in the Bayesian world something has changed: one door of the three is no longer relevant, so if your door has a one in three chance of hiding the car, then you need to make a Bayesian adjustment for the other door and assign it a new probability of two in three or 66%. Now, I know this seems completely counter-intuitive to most people, but this is the appeal of Bayes' rule; it carries out an error correction where we may not have originally found the need for one, as our perception of the scene can get in the way and make inferences that are, on a probabilistic basis, actually incorrect. If you change doors, you still may not win the car, but you will double your chances.

Predictive processing works in much the same way as Bayes' theorem; bottom-up sensory information is corrected by top-down predictive models, only where there is a disparity or a surprisal do our cognitive systems deploy the energy required to reconcile the anomaly. As each piece of new sensory information is processed and "corrected," so we gain more accurate information on what we are perceiving. If we only perceive the world on a top-down basis, then we must be incapable of seeing anything new or different. This is one of the basic points I am making in this book about the power of ancient drama to affect mental change in its audience members. But back to the Monty Hall problem for a moment. Bayes' rule proves that you should have switched doors to double your chance of winning the car, and yet in controlled studies using the same Monty Hall dilemma, most people stick with the original door when they should switch.[36] What I am arguing in this study is that Greek drama provided a place where people's usual perceptions of their world were provoked and challenged – and their minds might even be changed.

It is important to have at least a cursory understating of Bayes' theorem, as it has also been widely applied to machine learning and artificial intelligence and is currently having a profound impact on some areas of neuroscience, cognitive studies, and philosophy. However, Clark offers a wise word of caution on making the general assumption that the human brain is Bayesian by stating, "in its broadest form, this need only mean that the brain uses a generative model to compute its best guess about the world, given the current sensory evidence." As we have shown with Gregory's hollow mask experiment, the generative model can sometimes be incomplete or even totally wrong. The variegated conditions of perception lead to such cognitive approximations. Clark sums this up well by stating, "all optimal inference is Bayesian, but not all Bayesian inference is optimal."[37]

Probabilistic politics

One application of Bayes' theory that is of particular relevance to this study is the work of Christian List and Philip Pettit on the application of probability models on group agency and decision making. This is of particular importance when thinking about democratic and collective institutions such as the decisions made by juries, government committees, and centralized banking operations. They propose that in order to function properly, these kind of group agents need to attempt to form "true" beliefs about the world and reject "false" ones.[38] List and Pettit show how the so-called law of large numbers – the proposition that by pooling the diverse information of multiple individuals, the group can do better at tracking the truth than relying on an individual opinion – operates in a broadly Bayesian manner. List and Pettit also proposed that for certain deliberative tasks, particular group structures were important in truth tracking, such as (a) democratization, where the adoption of a majoritarian democratic structure outperforms dictatorial decision making; (b) decomposition, where a collective judgment on a broad decision is more accurate when divided into a series of smaller collective judgments examining different criteria; and (c) decentralization, where a group faced with complex judgment tasks involving several probabilities, subdividing different facets of the task to expert participants, can also yield positive results.[39]

Recently, this work on group agency has been profitably applied to ancient Athens in a study of the concept of *eikos* as probability in the Greek world. Wohl has described the ancient Greeks as "thinking in sophisticated ways about probability many centuries before its supposed emergence in the modern era," and the distributed nature of *eikos* was particularly important in fifth-century Athenian society.[40] Vincent Farenga has shown how the democratic citizens of Athens were adept at using the type of organizational probability models for group agency described by List and Pettit, based on *eikos*, which he translates as "what is likely or probably true." One such model is the "law of large numbers," which works to error-correct individual predisposed concepts of truth based on the group's active processing of new causalities, probabilities, or counterfactual situations. Athenian democratic institutions developed with this model in mind by "redefining the validity of what individuals contribute to group knowledge.[41] Farenga has traced the development of this cognitive turn from the age of Solon in the sixth century to the functions and institutions of the classical Athenian democracy, and proposes that *eikos* bridged the gap in the move from individual to pluralist agency "in effect opening citizens' minds to alternative senses of what was true or real."[42]

We can extend this same organizational democratic premise to the Athenian theatre, itself a state institution organized and funded by the democratic government of Athens. We know that around 500 BCE the annual Festival of Dionysos Eleuthereus, or the City Dionysia as it is often called by scholars, was substantially reorganized. The original festival dates back to at least the 530s and was probably established by Pisistratus as part of his overall plan to promote the city of Athens as the centralized place of governmental administration and cult practice in the region of Attica.[43] Pisistratus had come to power during a bitter series of conflicts between the various regions of Attica that can be broadly broken down into coastal communities, the people living in and around Athens, and the people from the more mountainous north. The original festival probably primarily featured Dithyrambs, choral songs to Dionysos sung by large male choruses. We do not know when tragedy was included or exactly how it developed, but we do have evidence for the performance of tragedy in the early fifth century and the addition of comedy in the middle part of the century. The Athenian theatre developed along with its new democratic institutions and came to play an important role in the mechanisms of Athenian group agency with its ability to open its audience's minds to new perspectives and possibilities. This link between the theatre and the democracy is key to understanding its power within the culture as a whole, and ultimately why Plato derided Athens as a *theatrocracy* (*Laws* 701a). This was a performance culture where the people had been politically empowered because their aesthetic judgment had been valued en masse by poets who sought success by eliciting their emotional responses. For all the efficiency and complexity of the Athenian collective decision-making organizations, they could be either enhanced or completely undermined by the effect of group emotionality, something at which the Athenian theatre excelled.

The archaeological evidence for the theatre in Athens also supports this close link between theatre and political life. Material remains indicate that around 530 a sanctuary to Dionysos was established on the southeast slope of the Acropolis, indicated by the remains of a small temple to the god found there. We will explore the form of the Athenian stage in Chapter Two, but at this time it is assumed that the performance space was simply a levelled area above the sanctuary with a wooden grandstand that took structural advantage of the hillside to accommodate the audience. Around 510–500 BCE, we see evidence that both the performance area (*orchestra*) and seating area or "seeing place" (*theatron*) were expanded and improved.[44] This corresponds to the date of the early democratic reforms of Cleisthenes in 508/7.[45] Cleisthenes was an Athenian aristocrat who came to prominence in Athenian politics after the tyrant Hippias was expelled from Athens.[46] This power vacuum was filled with a clash between the Spartan-backed aristocratic forces of Isagoras and the popular movement of Cleisthenes who promised widespread reform. Cleisthenes won out and reorganized the communities of Attica, the so-called *demes*, into 10 administrative groups (*phylai*). Each of these new Athenian tribes was made up of an equal number of demes from the old rival geographic divisions of coast, city, and highland regions. These 10 tribes were responsible for organizing military recruitment and checking the eligibility of their male members for military service and the right to vote in the new citizen assembly. Furthermore, the reforms stipulated that each tribe elect a *strategos* – the military commanders of Athens – and send 50 representatives to the Athenian legislative executive council, the *boule*. As well as these measures, Cleisthenes also stipulated that the 10 democratic tribes send 50 men and 50 boys to compete as dithyrambic choruses in the Festival of Dionysos.[47] Thus, the links between the decision-making institutions of the Athenian democracy, the military, and the theatre were mutually established and intertwined throughout the fifth century.

Further evidence of the close connections between drama and political decision-making comes from Herodotus's description of the reforms of Cleisthenes, our earliest account on the subject.[48] Herodotus tells a rather strange story about Cleisthenes' maternal grandfather, a man of the same name, who had taken over the rule of the northern Peloponnesian city of Sikyon in the early sixth century after expelling the neighboring Argives. Herodotus compares the reforms of the Athenian Cleisthenes with those of his grandfather and details how he replaced the existing local cults and festivals of the Argive hero Adrastus with that of his mythological enemy, the Theban Melanippus. The relics of Melanippus were established at the seat of government and the tragic choruses, which had hitherto honored Adrastus, and were re-dedicated to Dionysos. Therefore, the elder Cleisthenes' political ambitions were closely connected to the re-establishment of a performance festival for Dionysos. This is also the earliest textual mention of any performance of tragedy we have, which might also indicate that, at least in its earliest choral form, the genre was not necessarily a purely Athenian invention. More importantly, what we see here is the notion of the exploitation of performance festivals for political reasons and the establishment of new aetiologies that reinforce the cultural dispositions of the establishing regimes.

What the Cleisthenes story illustrates is the close connection between Athenian democracy as a group agency system and the institution of the theatre, which was financed by wealthy, and often aristocratic patrons, as a form of state taxation and cultural prestige.[49] Therefore, the political associations found within the narratives of the plays have rightfully been a frequent source of study.[50] Yet, we should also be aware of Athenian drama's relationship with the cult worship of Dionysos, a god known for shamanistic practices involving intoxication and ecstasy, and his association with liminality, epiphany, and "otherness."[51] The older festivals of Dionysos involved drunkenness, ribald behavior, and street processions, masking, extreme costumes, song and dance, obscenities, and the display of the phallus, all intended to create a transformational experience of both character and place. Thus, the worship of Dionysos was always a heady, intoxicating, emotional, and above all embodied experience. A large part of this book will be an exploration of how emotions were communicated by the masked, moving, and singing performers of ancient drama. In the next chapter we will examine the embodiment of ancient drama within the spatial environment it was created for and how this particular space on the southeast slope of the Acropolis at the Sanctuary of Dionysos Eleuthereus greatly enhanced Athenian drama's ability to open and change the minds of its audience and foster empathy. Before we do that, it is important to place predictive processing within the context of distributed cognition and explain what I mean by an "embodied experience."

Extending prediction

Up to now we have explored the predictive process as almost purely computational, as if the mind is somehow separated from the external world. Some proponents of predictive processing have even gone so far as to suggest that the concept places doubt on the veracity of the entire idea of distributed cognition – the theory that the human mind is extended beyond the brain and even the body out into the world.[52] I disagree. Theories of the extended mind, enactivism, cognitive embodiment, and embeddedness, the so-called 4E's, have proliferated since the publication of Andy Clark's and David Chalmers' seminal 1998 paper, "The Extended Mind". This was a persuasively argued thesis that humans use cognitive scaffolding placed out in the environment to think, and that the body and its multiple sensory abilities are part of a feedback loop that constantly flows between brain, body, and environment and back again.[53] Subsequently, Clark has postulated that embodied cognition actually plays a vital role in the mechanism of predictive processing, which is an integral part of what he describes as "a mobile embodied agent located in multiple empowering webs of material and social structure"; he goes on to say, "it is the predictive brain operating in rich bodily, social and technological context that ushers minds like ours into the material realm."

For many in the humanities, theories of embodied cognition have offered an attractive alternative to neuroscience, which some mistakenly believe offers only a deterministic view of the human mind.[54] To many people, modern neuroscience is represented by functional Magnetic Resonance Imaging studies (fMRI), which generated a great deal of excitement in the 2000s for offering the ability to moni-tor parts of the brain being "activated" by recording changing blood oxygen levels in different regions after a magnetic pulse. However, neuroscience is a far more complex and multifaceted discipline. On the whole, the broad field of neurosci-ence is far from deterministic. For example, the fields of cultural and social neuro-science offer compelling evidence for the impact of different cultural experiences on how the mind develops and functions.[55]

Yet there is also much disagreement among seemingly complementary, yet often fundamentally different, theories of distributed cognition. Broadly speak-ing, "embodiment" suggests that our mental functions are beyond just the brain and involve our whole bodies operating in the environment; "embeddedness" posits that the brain, body, and environment are equal aspects of the cognitive process; theories of the extended mind can encapsulate all of these views, but they mostly suggest that there is no barrier between brain, body, and environment. On the other hand, enactivists would suggest that there are indeed borders as fine grained as at the cellular level, and the distributed nature of the crossing of these "borders" involves the organism acting on and in its environment.[56]

The seminal thought experiment that Clark and Chalmers used to explain embodied cognition was the so-called parity principle, the idea that an extended method of cognition is equally effective as an interior one. They demonstrated this with two characters, Inger and Otto, who were both going to a museum. Inger relied on the internal mechanisms of her working memory to remember the route to the museum, whereas Otto, who had Alzheimer's, relied on notes he had writ-ten down in his notebook to guide him. Both methods helped Inger and Otto make it to the museum. Though Inger had seemingly used internal memory to map her way, the generative predictive model suggests that Inger's method is not purely internalist; she must have embodied the route at some point, either by walking it or spatially working it out by studying an external map. Likewise, Otto's method is not solely distributed out onto his notebook; his brain still processed the infor-mation inscribed there and used it to create the action needed to move toward the destination. Predictve processing, if correct, as construed by Clark, offers a reconcilement between internalist and externalist theories.

Predictive processing is therefore an integrated model of perception, cogni-tion, and action, suggesting that the brain is not the seat of inner "central cogni-tion" but "an organ for the environmentally situated control of action." Embodied agents operating in the sensory-rich environment help create fast-flowing deduc-tive inferences that "effortlessly span brain, body and world." In effect, Clark's theory integrates computational theories of mental processing supported by neu-roscience research with existing theories of distributed cognition, explaining how the constant flow of bottom-up sensory information collected by the body in the environment is dynamically compared to top-down models stored in the neural mechanisms of the brain.

Not everyone agrees. On the other side of the embodied predictive debate is Jakob Hohwy, who argues that although the body plays an important role in sense-gathering, the mind is inferentially secluded from the world.[57] In his view, the sensory information that provides the material from the world for prediction is separated from the system that is doing the predicting. In Bayesian terms, this is called a "Markov blanket" and acts as a border between the predictive system and the evidence it collects. Basically, in order to error-correct sensory signals, it must be possible for the predictive system to draw on inferences that are not a part of the original sensory evidence that has been collected. Clark resists Hohwy's representational or cognitive virtual reality notion of the mind and instead argues that action is the key to predictive processing, not representation. Thus, the mind is extended as the external environment offers possibilities for action and intervention. Clark also sees human-made environments as both a product of and a scaffolding for generative predictive processing, which includes tools, writing systems, machines, and systems of philosophy, astronomy, and logic.[58] I think Clark is right and that extended systems that provide "an endlessly empowering cascade of contexts" are more efficient at minimizing prediction error than the non-extended brain acting alone. I also think that theatre cannot be effective if we use Hohwy's "Markov blanket." In one sense it could be posited that it acts as a kind of cognitive proscenium arch by providing a frame or border to the stage action, but this is not how drama works. For an audience to be drawn into a story, they need to become absorbed by it and, as I hope to show, Greek drama had no such barrier and used its theatrical devices to enhance the affective and narrative connections between actor and audience. We will explore this process more deeply when we re-examine the tragic mask in Chapters Three and Four and productively apply extended mind concepts such as Malafouris's Material Engagement Theory (MET).[59] In general, this study accepts and applies Clark's unification of computational, Bayesian systems with models of extended cognition and his view that "the predictive brain is an action-orientated engagement machine—an enabling node in patterns of dense reciprocal exchange binding brain body, and world."[60] As we will see, embodiment, extension, and distributed cognition are essential for the effective and the affective communication of drama.

If action is key to effective prediction generation, then what facilitates that action and, getting back to the dark room problem, why would we not simply seek to mitigate surprisal by staying chained up in Plato's cave? The simple answer is that we need surprise, and my biggest argument against Hohwy's revival of cognitive internalism is that, whether we like it or not, we are beings that live in the world. There is no water, no food, no sexual partner, no stimulus, nothing to challenge us at all in the dark room, and in order to find these essentials we need to venture out into the environment. This is where our predictive system is most useful in managing risk and reward, and in doing so we rely on our affective systems to provide the physiological means we need to fight or fly (amongst the myriad other emotional responses we may have to any given situation). Surprises always elicit emotional responses as we quickly process the sensory information from our environment and then make predictions as to how to deal with what the world presents. The Greeks called this *eikos* and the Athenians based many of their social, political, and cultural institutions on the concept of probabilistic/predictive thinking, group action, and error correction. In sizing up these probabilities, people had to be open to receiving novel information and changing their minds – they had to explore the possibilities. In the next chapter we will explore how the theatre at the Sanctuary of Dionysos Eleuthereus provided the cognitive means to enhance such explorations into alterity.

Notes

1 Halliwell 1986: 76.
2 Halliwell 1986: 75.
3 Clark 2015: xvi.
4 Poe 2010: 166.
5 Friston 2010: 127–138.
6 Clark 2015: 264.
7 Clark 2015: 26.
8 Frith 2008: 111.
9 Clark 2015: 27.
10 Schwartenbeck et al. 2013: 710, cited in Clark 2015: 265.
11 Hohwy 2015: 15.
12 Kidd et al. 2012. See also, Schulz 2015: 42–43 and Téglás and Bonatti 2016: 227–236.
   As Zerubavel has noted, "the selective nature of our attention is evident not only in the organization of our sensory experience, but also in the remarkably similar organization of the way we think about and remember things." Zerubavel 2015: 5.
13 Brook 1996: 74.
14 Brook 1996: 75.
15 Rimer and Yamazaki 1984: 64–110.
16 Ortolani 1995: 121–122.
17 Yoshimi 2016: 15.
18 Dastur 2000: 178–189.
19 Dastur 2000: 182.
20 Dastur 2000: 178.
21 Merleau-Ponty 1964: 12–14.
22 Merleau-Ponty 1964: 7.
23 Clark 2015: 17–19.
24 Pagnoni and Guareschi 2015: 2.
25 Plutarch *Moralia* 998E; Nauck, *Trag. Graec. Frag.* 456.
26 Wohl 2014: 142.
27 Kirby 2012: 422.
28 Wohl 2014: 142.
29 Kraus 2007: 5.
30 Hacking 2006.
31 Hoffman 2008: 1–29.
32 Kraus 2006: 129–150.
33 E-Chalk, The Rotating Mask Illusion. www.youtube.com/watch?v=sKa0eaKsdA0. Accessed September 22nd 2016; www.youtube.com/watch?v=QbKw0_v2clo. Accessed September 22nd 2016.
34 Clark 2015: 51.
35 McGrayne 2011: 89–136.
36 Granberg and Brown 1995: 711–723.
37 Clark 2015: 303.
38 List and Pettit 2011.
39 List and Pettit 2011: 86–100.
40 Wohl 2014: 5.

41 Farenga 2014: 85.
42 Farenga 2014: 85.
43 Meineck 2012: 20–22.
44 See Connor 1990, who argues for a date of around 500 BCE; however, Cartledge 1997: 23–24 is probably right in placing the date of the foundation of the City Dionysia at around 530 BCE. See also Sourvinou-Inwood 1994: 275–277 and Rhodes 2003: 104–119.
45 For the date of the reforms of Cleisthenes, see Wilson 2000: 12–19.
46 Herodotus 5.66; *Aristotle Ath.* Pol. 21.2, Pol. 1275b 34–40, 1319b 19–27; Plutarch, *Pericles* 3.2–3.
47 Pritchard 2004: 208–228; Zimmerman 1993: 18–20; Wilson 2003; Fisher 1998.
48 Herodotus 5.67.1–68.2.
49 On the institution of *Choregia*, see Wilson 2000.
50 For example, Carter 2011; Pirro 2011; Sommerstein 2014: 291–305.
51 Seaford 2009: 25–38.
52 Clark and Chalmers 1998: 7–19.
53 Clark 2015: xvi.
54 Noë 2009.
55 Northoff 2016; Harmon-Jones and Inzlicht 2016; Cacioppo 2016; Hyde et al. 2015; Baron-Cohen et al. 2013; Chiao et al. 2013; Ames and Fiske 2010.
56 Kiverstein and Clark 2009: 1–7; Menary 2010: 459–463.
57 Hohwy 2013; de Bruin and Michael 2016
58 Clark 2013: 195.
59 Malafouris 2013.
60 Clark 2015: xvi.

Bibliography

Ames, D.L. and Fiske, S.T. 2010. "Cultural neuroscience" in *Asian Journal of Social Psychology* 13.2: 72–82.

Baron-Cohen, S., Lombardo, M. and Tager, H. (Eds.). 2013. *Understanding other minds: Perspectives from developmental social neuroscience*. Oxford: Oxford University Press.

Brook, P. 1996. *The empty space: A book about the theatre: Deadly, holy, rough, immediate*. New York: Simon and Schuster.

Bruin, L. de and Michael, J. 2016. "Prediction error minimization: Implications for embodied cognition and the extended mind hypothesis" in *Brain and Cognition* 112: 58–63.

Cacioppo, J.T. 2016. *Social neuroscience*. Cambridge, MA: MIT Press.

Carter, D.M. 2011. *Why Athens? A reappraisal of tragic politics*. Oxford: Oxford University Press.

Cartledge, P. 1997. "'Deep plays': Theatre as process in Greek civic life" in Easterling, P.E. (Ed.), *The Cambridge companion to Greek tragedy*. Cambridge: Cambridge University Press: 3–35.

Chiao, J.Y., Cheon, B.K., Pornpattananangkul, N., Mrazek, A.J. and Blizinsky, K.D. 2013. "Cultural neuroscience" in Gelfand, M., Chiu, C. and Hong, Y. (Eds.), *Advances in culture and psychology 4.1*. Oxford: Oxford University Press: 1–77.

Clark, A. 2013. "Whatever next? Predictive brains, situated agents, and the future of cognitive science" in *Behavioral and Brain Sciences* 36.3: 181–204.

Clark, A. 2015. *Surfing uncertainty: Prediction, action, and the embodied mind*. Oxford: Oxford University Press.

Clark, A. and Chalmers, D. 1998. "The extended mind" in *Analysis* 58.1: 7–19.

Connor, W.R. 1990. "City dionysia and the Athenian democracy" in Connor, W.R. (Ed.), *Aspects of Athenian democracy* (Vol. 11). Copenhagen, Denmark: Museum Tusculanum Press, University of Copenhagen: 7–32.

Dastur, F. 2000. "Phenomenology of the event: Waiting and surprise" in *Hypatia* 15.4: 178–189.

Farenga, V. 2014. "Open and speak your mind" in Wohl, V. (Ed.), *Probabilities, hypotheticals, and counterfactuals in ancient Greek thought*. Cambridge: Cambridge University Press: 84–100.

Fisher, N. 1998. "Gymnasia and the democratic values of leisure" in Cartledge, P., Millet, P. and von Reden, S. (Eds.), *Kosmos: Essays in order, conflict, and community in classical Athens*. Cambridge: Cambridge University Press: 84–104.

Friston, K. 2010. "The free-energy principle: A unified brain theory?" in *Nature Reviews Neuroscience* 11: 127–138.

Frith, C. 2008. *Making up the mind: How the brain creates our mental world*. London: Blackwell.

Granberg, D. and Brown, T.A. 1995. "The Monty Hall dilemma" in *Personality and Social Psychology Bulletin* 21.7: 711–723.

Hacking, I. 2006. *The emergence of probability: A philosophical study of early ideas about probability, induction and statistical inference*. Cambridge: Cambridge University Press.

Halliwell, S. 1986. *Aristotle's Poetics*. Chicago: University of Chicago Press.

Harmon-Jones, E. and Inzlicht, M. (Eds.). 2016. "A brief overview of social neuroscience" in Harmon-Jones, E. and Inzlicht, M. (Eds.), *Social neuroscience: Biological approaches to social psychology*. London: Routledge: 1–9.

Hoffman, D.C. 2008. "Concerning eikos: Social expectation and verisimilitude in early Attic rhetoric" in *Rhetorica: A Journal of the History of Rhetoric* 26.1: 1–29.

Hohwy, J. 2013. *The predictive mind*. Oxford: Oxford University Press.

Hohwy, J. 2015. "The neural organ explains the mind" in Metzinger, T. and Windt, J.M. (Eds.), *Open MIND: 19(T)*. Frankfurt am Main, Germany: MIND Group: 1–22.

Hyde, L.W., Tompson, S., Creswell, J.D. and Falk, E.B. 2015. "Cultural neuroscience: New directions as the field matures" in *Culture and Brain* 3.2: 75–92.

Kidd, C., Piantadosi, S.T. and Aslin, R.N. 2012. "The Goldilocks effect: Human infants allocate attention to visual sequences that are neither too simple nor too complex" in *PloS One* 7.5: e36399.

Kirby, J. 2012. "Aristotle on Sophocles" in Ormand, K. (Ed.), *A companion to Sophocles*. Malden, MA: Blackwell: 411–423.

Kiverstein, J. and Clark, A. 2009. "Introduction: Mind embodied, embedded, enacted: One church or many?" in *Topoi* 28.1: 1–7.

Kraus, M. 2006. "Nothing to do with truth? Eikos in early Greek rhetoric and philosophy" in *Papers on Rhetoric* 7: 129–150.

Kraus, M. 2007. "Early Greek probability arguments and common ground in dissensus" in *Scholarship and UWindsor OSSA7*. Windsor, ON: University of Windsor. http://scholar.uwindsor.ca/ossaarchive/OSSA7/papersandcommentaries/92/. Accessed November 7th 2016.

List, C. and Pettit, P. 2011. *Group agency: The possibility, design, and status of corporate agents*. Oxford: Oxford University Press.

McGrayne, S.B. 2011. *The theory that would not die: How Bayes' rule cracked the enigma code, hunted down Russian submarines, & emerged triumphant from two centuries of controversy*. New Haven, CT: Yale University Press.

Malafouris, L. 2013. *How things shape the mind.* Cambridge, MA: MIT Press.

Meineck, P. 2012. "The embodied space: Performance and visual cognition at the fifth century Athenian theatre" in *New England Classical Journal* 39: 3–46.

Menary, R. 2010. "Introduction to the special issue on 4E cognition" in *Phenomenology and the Cognitive Sciences* 9.4: 459–463.

Merleau-Ponty, M. 1964. *The primacy of perception: And other essays on phenomenological psychology, the philosophy of art, history, and politics.* Evanston, IL: Northwestern University Press.

Noë, A. 2009. *Out of our heads: Why you are not your brain, and other lessons from the biology of consciousness.* New York: Palgrave Macmillan.

Northoff, G. 2016. "Cultural neuroscience and neurophilosophy: Does the neural code allow for the brain's enculturation?" in Chiao, J., Li, S.C. and Turner, R. (Eds.), *The Oxford handbook of cultural neuroscience.* Oxford: Oxford University Press: 21–39.

Ortolani, B. 1995. *The Japanese theatre: From shamanistic ritual to contemporary pluralism.* Princeton, NJ: Princeton University Press.

Pagnoni, G. and Guareschi, F.T. 2015. "Remembrance of things to come: A conversation between Zen and neuroscience on the predictive nature of the mind" in *Mindfulness* 8.27: 1–11.

Pirro, R.C. 2011. *The politics of tragedy and democratic citizenship.* London/New York: Bloomsbury.

Poe, M.T. 2010. *A history of communications: Media and society from the evolution of speech to the Internet.* Cambridge: Cambridge University Press.

Pritchard, D. 2004. "Kleisthenes, participation, and the dithyrambic contests of late archaic and classical Athens" in *Phoenix* 58.3: 208–228.

Rhodes, P.J. 2003. "Nothing to do with democracy: Athenian drama and the polis" in *The Journal of Hellenic Studies* 123: 104–119.

Rimer, J.T. and Yamazaki, M. 1984. *On the art of the no drama: The major treatises of Zeami.* Princeton, NJ: Princeton University Press.

Schulz, L. 2015. "Infants explore the unexpected" in *Science* 348.6230: 42–43.

Schwartenbeck, P., FitzGerald, T., Dolan, R. and Friston, K. 2013. "Exploration, novelty, surprise, and free energy minimization" in *Frontiers in Psychology* 4: 710.

Seaford, R. 2009. "Tragedy and dionysus" in Bushnell, R. (Ed.), *A companion to tragedy.* Malden, MA: Blackwell: 25–38.

Sommerstein, A.H. 2014. "The politics of Greek comedy" in Revermann, M. (Ed.), *The Cambridge companion to Greek comedy.* Cambridge: Cambridge University Press: 291–305.

Sourvinou-Inwood, C. 1994. "Something to do with Athens: Tragedy and Ritual" in Osborne, R. and Hornblower, S. (Eds.), *Ritual, finance, politics: Athenian democratic accounts presented to David Lewis.* Oxford: Oxford University Press: 269–290.

Téglás, E. and Bonatti, L.L. 2016. "Infants anticipate probabilistic but not deterministic outcomes" in *Cognition* 157: 227–236.

Wilson, P. 2000. *The Athenian institution of the Khoregia: The chorus, the city and the stage.* Cambridge: Cambridge University Press.

Wilson, P. 2003. "The politics of dance: Dithyrambic contest and social order in ancient Greece" in Phillips D.J. and Pritchard, D. (Eds.), *Sport and festival in the ancient Greek world.* Swansea, Wales: Classical Press of Wales: 163–196.

Wohl, V. (Ed.). 2014. *Probabilities, hypotheticals, and counterfactuals in ancient Greek thought.* Cambridge: Cambridge University Press.

Yoshimi, J. 2016. *Husserlian phenomenology: A unifying interpretation*. New York: Springer International.

Zerubavel, E. 2015. *Hidden in plain sight: The social structure of irrelevance*. Oxford: Oxford University Press.

Zimmerman, B. 1993. "Das Lied der Polis; Zur Geschichte des Dithyrambos" in Sommerstein, A. H., Halliwell, S., Henderson, J. and Zimmermann, B. (Eds.), *Tragedy, comedy and the polis*. Bari, Levanti Editori: 39–54.

2 *Opsis*

The embodied view

Imagine for a moment trying to drive without the means to anticipate the future and make constant self-correcting predictions. As we drive, we know from prior learning that by turning the steering wheel the car will go in a certain direction, but to use Merleau-Ponty's example of the unseen back of the lamp, we cannot know what is around that corner. We predict that the road will continue and that it will be clear, but what if instead we suddenly encounter a large truck on fire and stopped in the middle of the road. We are shocked and surprised and must immediately perform an error correction in order not to slam into the back of the truck and perish. At this moment our bodily autonomic systems kick into high gear, our heart rate increases, and we are flushed with neurochemicals such as dopamine and adrenaline that quickly force energy to our muscular and cardiovascular systems. Our skin temperature changes as a result, and our blood pressure increases as our body quickly calls for more oxygenated blood. Our breathing rate increases to provide that oxygen, and our pupils react as we make incredibly quick predictions about how to either stop the car or swerve to avoid slamming into the truck. If we have practiced these kinds of evasive moves before and learned prior action orientation representations, then we stand a better chance of survival. However, if it's a completely new situation, then we do our best, but sometimes the surprise overwhelms us or the environmental situation cannot be surmounted. This is a real possibility every time we drive, yet we choose to drive believing in the probability that it won't happen. Even if you don't drive, try just crossing the road in any busy city without anticipating the traffic conditions and constantly updating the sensory information as you cross. If you don't do so and just head straight into the road without mentally calculating the rate of speed of the oncoming traffic and how much time you have to cross, you stand a very good chance of being struck. It is not the darkened room that is the key to mitigating surprise, it is anticipation, and a key ingredient to this vitally important cognitive process is the neurotransmitter dopamine and its relationship to prediction error correction, action, and abstract thought.

In this chapter I want to advance the theory of prediction processing in practical terms and show how the form of the theatre itself greatly enhanced these cognitive processes on a neurochemical basis. To this end, I will describe the fifth-century theatre space with the benefit of new archaeological evidence that is challenging much of what we thought we knew about the classical stage, and re-evaluate what it meant to stage these works in this particular "open-air" theatre. We will consider how having the sky as the most dominant sight in the visual realm increased the brain's production of dopamine, the neurotransmitter that is necessary for memory recall, movement, imagination, prediction, and abstract thought.

This chapter also seeks to re-examine Aristotle's concept of *opsis*, most frequently and rather derogatorily translated as "spectacle." However, I think Aristotle was fully aware of the theatrical power of visuality, and he cites *opsis* as the single mode of realizing tragic mimesis and the way in which it was ordered (*kosmos*). For Aristotle, *opsis* encompassed the other five elements of tragedy – speech (*lexis*) and song (*melopoiia*) as the "means employed"; and narrative (*mythos*), character (*ethos*), and intention (*dianoia*) as the "object represented" (1449b32–38) – and *opsis* conveyed them all. Yet later in *Poetics* (1450b16–20), Aristotle places *opsis* at the bottom of his list of tragic elements and describes it as *atekhnos*. This has been variously translated as "very inartistic" or "artless," but the word also can mean "intrinsic" or something that cannot be made by a human.[1] In the same place, Aristotle calls *opsis*, *psychagōgikos*, which is often translated as "attractive" but has the sense of something that "enthralls the soul." Plato uses the same term in *Minos* (321a4) to describe tragedy as the most "soul-enthralling" branch of poetry. I want to investigate here what was intrinsic and soul-enthralling about dramatic visuality and in what way it lay beyond the control of the poet.

Environment and action

The term *psychagōgikos* contains the sense of something that moves or acts upon the person – in cognitive terms it might be described as an affordance. An affordance is the relationship between the world, or an object in it, and an actor (any organism) that by stimulation affords the opportunity for action. J. J. Gibson coined the term to describe the action possibilities that are latent in the environment, and Rush Rehm has applied Gibson's work on affordances to the Greek theatre:[2]

> Applied to a given tragedy, the idea of affordances can play into the changing dynamics of the drama. The humiliated Ajax in Sophocles' play, for example, no longer perceives the sword of Hector as a status gift befitting a war hero. Instead, the sword reverts to its original function, affording Ajax the means to fall by an enemy weapon and so assume the appearance of a hero's death in a world that has lost its heroic values.[3]

The theatre is an environment of mimetic affordances, and the successful reception of the relationships created between actor, prop, stage set, entrance and exits, performance space, and audience is essential for the act of theatre to be comprehensible. The generation of dramatic affordances that are organized within a predictive mimetic structure allows theatre to work. What I am getting at here has been much better expressed by Bharata Muni in the *Natyasastra*, his early treatise on Sanskrit drama where he describes *Rasa*, the perceptual blend that makes up a performance:

> Rasa is the result of the bhavas – feelings intended by the poet conveyed by words, the body, expressions. Bhavas are described as being made up of external affects (vibhavas), automatic responses (anubhavas), conscious responses (uyabhicaribhavas) and the master, the sthayibhavas, which is the total effect of this reaction.[4]

As Richard Schechner has written, there is no *Natya* (performance) without *Rasa* (tastes): "Rasa is flavor, taste, the sensation one gets when food is perceived, brought within reach, touched, taken into the mouth, chewed, mixed, savored, and swallowed."[5] Rasa is anticipation, reward, and satisfaction, as Bharata has it, "without *rasa* no purpose is fulfilled."[6] In the human brain and body, anticipation, reward, and the resulting satisfaction are in large part facilitated by the neurotransmitter dopamine.

In the popular imagination, dopamine has been primarily associated with the reward and pleasure centers of the brain, but its pathologies are far more complex, and this vital chemical neurotransmitter has been linked to cortical excitability and attention deficit, motor control, working memory, schizophrenia, reinforcement learning and addiction, executive function, exploration, and reward prediction.[7] In terms of affordances and predictive processing, dopamine regulates the affordance of sensory cues that create motor behavior. According to Friston et al., "the emergent role of dopamine is to report the precision of perceptual cues that portent a predictable sequence of sensorimotor events – inferring that an object has affordance necessarily entails an action." Dopamine may be central in this inference and implicit action selection.

To put this more simply, dopamine is the chemical element that helps facilitate the act of predictive error correction, and its links to working memory, motor control, and reward are not at all unrelated. Dopamine might best be described as a stimulator of human possibilities. Dopamine helps us to cognitively time travel, to think back into our perceptual pasts, and also to imagine future scenarios: dopamine allows us to *act* – it is, in many ways, akin to Anaximenes' *kerdos* (impulse), one of his three parts of probabilistic thought (*eikos*) along with *ethos* (cultural practice) and *pathos* (emotion).[8] Although dopamine is one chemical part in the complex neurobiological processes of perception, prediction, and action, it is nevertheless an essential element in the generation of imagination that makes theatre possible. Here I will suggest that the physical setting of the Theatre of Dionysos in Athens was an extremely dopamine-rich environment and how this helped facilitate the dramas of alternative realities that the Athenians experienced there.

Phantasia theatre

For evidence that the Greeks were also aware of the cognitive powers of anticipation and reward, we can turn again to Aristotle. In *de Anima* he explains the concept of *phantasia* (imagination) as linked to the human perceptual process and how it tends to evaluate and seek what is pleasurable and good. On this subject, Jennifer Moss has proposed that Aristotle's idea of the motivating power of pleasure outweighs reason or logic in human decision making. When we examine how he defines *phantasia* and the processing of pleasure, we find an uncanny similarity with the way in which modern neuroscience has revealed the cognitive operations of dopamine.[9] For example, Moss shows how Aristotle equates *phantasia* with the act of movement and that movement is always goal directed (*de Anima* 432b15). Moss explains that creatures apprehend objects as goals through imagination and that goal-directed behavior involves some cognitive reward reinforcement or feelings of pleasure, which we might call anticipation. Thus, *phantasia* is responsible for associating certain objects with pleasure or reward.[10] If we recast what Aristotle calls *phantasia* with the neural attributes of dopamine, we can start to see how he is articulating the relationship between working memory and prediction (imagination), sensory information (perception), anticipation (pleasure), and action (movement). We might even go so far as to equate Aristotle's *phantasia* with predictive processing in that *phantasia* is posited as more than just perception. As Moss points out, even translating the word as "imagination" is misleading; a better translation might be "what things appear to be" or "made apparent." This might go some way to explaining why Aristotle applied *phantasia* to so many cognitive processes that at first sight seem unrelated until they are viewed within a predictive context.

Aristotle even goes so far as to apply *phantasia* directly to occurrences of perceptual error. In *de Anima* he offers a complex description of *phantasia* that is fascinatingly most concerned with a fluid process of error correction that he likens to a kind of movement: "imagination must be a movement produces by sensation actively operating" (429a32–34). He chooses as his example the perception of the size of the sun in the sky: it appears sensorially to measure only a foot across, but we predict, based on prior knowledge, that it is greater than the world. Therefore, our perceptual system is constantly corrected and updated (428a–429a10). Moss calls this aspect of *phantasia* "a capacity for making present to the mind something one has perceived before" and that "through bare perception we can become aware of an object, but only through *phantasia* can we apprehend it as something we might want to pursue or avoid."[11] Thus, like the functions stimulated by dopamine, *phantasia* moves us to action (*de Anima* 428b25–429a2). Moss calls this "envisioning prospects" and posits that it is not just perception and a memorialized representation that spurs us to action but a more complex affective tagging that also involves prediction (e.g., the contemplation of the quenching of our thirst or the protectiveness of shelter).

We find similar ideas in Plato. In *Philebus* he has Socrates say that sensations ("pleasures and pains") that "belong to the soul alone" can come before bodily sensations so that we have the sensation of anticipation (39d) and "we are always filled with hopes all our lives" (32c1). Furthermore, in a detailed discussion on consciousness in *Sophist*, Plato writes that what is brought about by sensation is *phantasia*, and he describes it as "a mixture of sensation and opinion," noting that some of these must be errors (264b).

Anticipation and action have been shown to be products of the predictive coding properties of dopamine. A groundbreaking 1998 Cambridge study by Wolfram Schultz brought this to light. This research project trained a group of monkeys to pull a lever to receive a small edible treat.[12] A light signal was added, which would indicate when the treat was available. No light, no treat, even if the monkey pulled the lever. His initial study found that the dopamine levels were higher at the light signal than when the treat was received. Dopamine spiked for anticipation more than for receipt of the actual reward. But Schultz's studies became all the more remarkable when he recalibrated the same experiment. This time when the light went on, the monkey received the treat only 50% of the time at averaged randomized intervals. One would expect the dopamine levels to decrease or at least stay the same, but instead they increased substantially. This insertion of chance or probability significantly increased the dopamine levels and therefore the anticipation. Then the test was recalibrated again, first to a 25/75 ratio and then to a 75/25 ratio. The first produced the lowest dopamine rates as there was not enough prediction reliability to sustain attention and expectancy, and the second put the dopamine levels back up to the 100% levels as the frequency of the treat being dispensed was more predictable.

This research showed that dopamine is connected with anticipation in the ordering of action. The light goes on, the monkey anticipates, and then moves to pull the lever. But just like Aristotle's comments on the right amount and type of unexpectedness in drama, dopamine operates in a similar manner, being at its most excited when symmetries of cause and effect enhance predictive success. This research goes far in helping to alleviate the darkened room problem: although the sensorial deprivation afforded there is most conducive to the free energy principle, in order to survive we need to enact upon and be active within our environments, to locate the nourishment we need to exist. This involves risk and reward, thus anticipation encoded by dopamine through predictive learning is essential. As Friston et al. claim, dopamine balances bottom-up sensory information and top-down prior beliefs when making hierarchical inferences (predictions) about cues that have affordance.[13] Furthermore, it encodes the precision of beliefs about alternate actions and controls the subsequent behavior.[14] Thus, dopamine provides the neurochemical means by which top-down predictions meet bottom-up signals and create a motivation for attention and action.[15] Dopamine is the lubricant of our imaginations – the fuel of what Plato and Aristotle called *phantasia* – the human ability to be motivated to explore possibilities; and the Theatre of Dionysos was a dopamine-inducing environment.

The theatre is nothing if not the realm of *phantasia*, and today, in the West, most theatre is created for a darkened room, its possibilities and probabilities displayed by means of representation – Aristotle's *mimesis*. In many respects, this form of interior theatre has prevailed over the past century and has colored the way in which many have come to view Greek drama. This is despite attempts, mostly from the avant-garde, to break theatre out of its modern relationship with the so-called fourth wall – the audience, gathered inside to watch theatre performed in front of them in some form of a proscenium arch space. There is also the prevailing notion of the theatre as a forum for the performance of a text and the enactment of the author's ideas. While this is certainly true of both modern and ancient drama, we must also be acutely aware of the experiential nature of ancient drama, with its roots in the shamanistic rituals of Dionysos, dance, song, masked performance, and the incorporation of the environment through procession, sacrificial ritual, and the setting of the plays in the open-air theatre with its spectacular views of city, landscape, and sky. Artaud railed against the modern theatre's lack of experientially, calling it a "theatre of dialogue" and asking "how is it that Occidental (Western) theatre does not see theatre under any other aspect?" (37). This is not at all true today, where experimental, physical, and environmental theatre works, to name but a few, have sought to move Western theatre in new experiential directions. Although Artaud may have been extreme in his rejection of dialogue as a theatrical form, he was right in trying to situate text as an enacted event in a wider context of the overall sensory experience of live drama:

> I say that the stage is a concrete physical space which asks to be filled, and to be given its own concrete language to speak. I say that this concrete language, intended for the senses and independent of speech, has first to satisfy the senses, that there is a poetry of the senses as there is a poetry of language, and that this concrete physical language to which I refer is truly theatrical only to the degree that the thoughts it expresses are beyond the reach of the spoken language.[16]

What does Artaud mean by "beyond the reach of spoken language"? In terms of the ancient theatre, we might think of the idea of *phantasia* and the sensory totality of ancient performance, which we will discuss further when we explore masks, movement, and music. But we might also consider the Athenian's theatre relationship to Dionysos, a god of epiphanies, ever arriving and surprising and bringing with him cult practices that encourage behaviors that take the participants out of their ordinary and everyday existence into an altered state. This was achieved via alcohol, music, procession, and the kind of cognitive absorption that I am arguing ancient drama could induce. In many ways, Dionysos is beyond the reach of spoken language, and as Andrew Ford has shown, and I discuss in greater detail in Chapter Six, he is also a god of unintelligibility often expressed at the height of emotions. Dionysos was the god of the altered state, of ecstasy – the standing outside of oneself – and a god of physical places of liminality – the mountains, the wild countryside, faraway foreign lands on the edge of the known world, borders, and the place where earth and sky meet. The theatre of Dionysos expressed these aspects of its god, in terms of its location on the southeastern slope of the Acropolis affording panoramic views of the old southern city, the low-lying mountains of the Attic countryside, and, most importantly, the sky.

Placing Athenian drama within its original performative environmental context is essential if we are going to try to understand how it functioned and what made it such a compelling emotional, cultural, and political force. In order to do this, let's first outline what we currently know about the fifth-century theatre based on the

latest archaeological research, which in many ways is questioning a lot of prevalent ideas about the form and function of the Greek theatrical performance space.

The seeing place of Dionysos

When most people, including many scholars of both theatre and classics, imagine
the theatre where the plays of Aeschylus, Sophocles, Euripides, and Aristophanes
were performed, many will still describe a large stone amphitheatre, with a circular *orchestra*, some 16,000 seats arranged in steep rakes around two-thirds of the
circle, with a stage and scenery at one end. But what they are envisioning is the
beautiful theatre at Epidauros, where performances are still held, or the Athenian
stone theatre of Lycurgus, the remains of which and later Roman renovations can
be seen today. These theatres date to the second half of the fourth century BCE,
some 80–100 years later than the time when the majority of the plays we have
were performed (the main exceptions are the tragedy *Rhesus*, which is wrongly,
in my opinion, attributed to Euripides, and the comic plays of Menander). In fact,
what we now know about the *theatron* ("viewing stand") erected over the Sanctuary of Dionysos Eleuthereus in Athens is that it was far smaller and simpler than
the traditional concept of the classical theatre.

New appraisals of epigraphic, archaeological, and literary evidence have seriously challenged the prevailing view that the theatre was a large (16,000 seats
plus) venue that bordered a circular playing space. Instead, the evidence indicates that in the fifth century BCE the theatre was a temporary wooden structure
of around 6,000 seats with a rectilinear, not circular orchestra.[17] These arguments are by now well known to specialists in the field, with the debate about
the shape of the *orchestra* dating back to the work of Carlo Anti in 1947 and
Elizabeth Gebhard in 1974 and forming an important part of the work of David
Wiles in 1997.[18] Despite this knowledge, most people still imagine the Greek
theatre space as a vast stone, beautifully proportioned edifice – an "inward facing circle"[19] where the *demos* could all gather to watch productions and each
other, with the dramas performed there part of a great self-conscious display of
civic ideology.[20] Now, in the light of the new evidence, this view needs to be
reconsidered.

Most have regarded the Athenian performance space in modern terms as a play-
house – a structure built to house audiences watching plays, rather than a viewing
place (*theatron*) erected at the terminus of a great ritual procession, for spectators
to watch performances of dance, music, and lyrics in an open-air environment.
This setting was not the design of an architectural plan but a development of an
existing choral- and processional-based performance culture, combined with the
foundation of a new festival – the City Dionysia, around 530 BCE, which established a sanctuary to Dionysos as an environmental *theatron* – a "seeing place."
Let us examine the idea of the "seeing place" from a cognitive perspective and
explore one of the most basic premises of the staging of ancient drama – that
the venue was in the open air. First, what exactly did the spectators seated in the
theatron see? Not just the plays staged before the sanctuary but also the sanctuary

itself (animal sacrifices were carried out during the performances), the southern city beyond that, and then the Attic countryside. But the most prominent element in their visual field was the vast expanse of the sky. One can get an impression of how the Acropolis dominated the surrounding Attic plain by looking at photos taken in the 1860s before the modern development of Athens (Figure 2.1).

The most recent archaeological survey of the site has found post holes dating from the fifth century with the imprint of wood grain preserved in the soil. This reveals two important things: the limit of the fifth-century *theatron*, and that the *ikria* or seats may not have been temporary, despite some earlier claims based on later evidence.[21] Although it is possible that the planking used for seating was removed and the structural uprights that supported them stayed in place, as wood was an expensive commodity in antiquity and warped in the sun and winter if left exposed. In the fourth century, we know of specialist producers who toured seating and entered into contracts with the state to provide festival grandstands.[22] As for the capacity of the space, based on this evidence, several scholars now estimate that around 6,000 people could be accommodated.[23] The figure of 17,000 seats originally proposed by Pickard-Cambridge in the mid-20th century has been influential and is still cited in many books on ancient drama,[24] yet even though this was his estimate for the fourth-century Lycurgan theatre, it became widely and mistakenly adopted as the size of the fifth-century *theatron*.[25] The capacity of the *theatron* is an important factor in considering the social function of the theatre in that a large theatre capable of accommodating many thousands of citizens is a

Figure 2.1 The Athenian Acropolis from the southeast, 1868–1875. Félix Bonfils, Brünnow Papers, Manuscript Division, Princeton University.

very different cultural event than one that seats 6,000, many of them non-Athenian visitors. Yet as I will explain, this does not necessarily mean we should not still consider the Theatre of Dionysos a "democratic" space.

The archaeological findings offer evidence for a close relationship between the theatre and politics. The venue was most likely instituted as a sanctuary to host the new state-funded City Dionysia around 530. It was enlarged and improved around the time of the reforms of Cleisthenes, c. 510–500 BCE, which ushered in new democratic reforms, and then again around 430 BCE. This coincided with the Athenian building program that reinforced the role of Athens as head of the Delian League, a military alliance against Persian and Spartan influence.[26] Carter has suggested that this theatre was less about promoting the civic ideology of the Athenian demos but rather more about the promotion of Athens as a benign imperial protector.[27] However, these two positions were not unrelated in the fifth century. A capacity of 6,000 is still a very large venue by theatrical standards (most Broadway and West End houses are between 1,000–2,000 seats), and it was still one of the largest public spaces ever developed in fifth-century Athens. Importantly, the theatre held the same amount of people as the Pnyx, where the assembly met to vote on Athenian policy, a venue that must be regarded as a "democratic space."[28]

The evidence also suggests that there was no circular orchestra at this point and that the *theatron* may have been predominantly frontal, as suggested by the remains of sixth- and fifth-century spaces, including the theatres found at Argos and Lavrio.[29] Apart from the front row of stone seats and improvements to the *skene* added in the 430s, the theatre space was a large wooden grandstand overlooking a dancing place before a sanctuary, on the slope of the most sacred site in Attica – the Acropolis – and commanding the most incredible and affecting view. What this new evidence indicates is that instead of thinking about the classical Athenian theatre in terms of an architectural space, we should instead see it as an environmental space, whose playing area reflected the origins of drama in processional display and chorality, and whose site was developed to deliberately take advantage of the panoramic views and place its audiences in a direct and powerful relationship to the sky.[30]

The later form of the Greek theatre, such as the great stone theatres at Epidauros and of Lycurgus in Athens, continued to reflect these environmental elements, such as a panoramic view and the processional and choral influences on drama. For example, the wing entrances, *eisodoi* (side roads), are a vestige of the processional route where from time to time the chorus would halt to perform. This created a dynamic and fluid theatre of movement where what was "onstage" and what was "offstage" was more ambiguous and dynamic. Before the use of the *skene* (scene building), with its central doorway, performers came into the audience's view at different times and action flowed continuously.[31] As we shall see in Chapter Four, in this kind of space Greek drama was not viewed as being made up of scenes interrupted by choruses, but instead of choruses flowing in and out of scenes, informing the narratives and moods of each – a dynamic movement-based theatre.

One misunderstanding of late ancient evidence found in the large Byzantine encyclopedia known as the *Suda* is that the theatre was originally located in the Agora and only moved to the Acropolis when the stands collapsed. However, these entries have been mistakenly conflated by scholars who had supposed that any reference to *ikria* – wooden benches – could not mean the Theatre of Dionysos, which they assumed always had stone seating. *Ikria* were not exclusive to the theatre but used to accommodate crowds for all kinds of public events at a variety of locations.[32] Furthermore, the archaeological evidence for the site on the Acropolis slopes dates from around 530 BCE, indicating that the location was deliberately chosen for the festival when it was established at that time. Thus, we should not assume that the theatre was moved from an earlier site in the Agora, but instead deliberately established when the City Dionysia was inaugurated. This site was purposely chosen because it offered such panoramic views intended to represent liminal distal space – the realm of Dionysos. This visual array was a crucial part of the cult worship of Dionysos and therefore affected the development of the dramatic performances that were produced there. What I am suggesting is that rather than focusing its audiences' attention away from the scenery and sky, as most modern theatre tries to do, this view was actually an integral part of the way in which ancient Greek drama was originally experienced.

There is further evidence that this new sanctuary was intended to represent liminal distant space in its name – the Sanctuary of Dionysos Eleuthereus. The town of Eleutherae was located on the border between Attica and Boeotia some 27 miles to the north. Scholars have attempted to explain the connection with Eleutherae, but the very nature of Athenian aetiological myths, whereby the origins of a cultic or civic event are often deliberately obfuscated, means that any precise explanation is bound to fail.[33] A sense of sacredness and age-old practice is created by attaching myths to certain visible physical locations and local customary practices and is often enacted by means of performance.[34] The name "Dionysos Eleuthereus" has led some to believe that it is a political reflection of the annexation of the previously Theban town of Eleutherai into Attica.[35] However, it is far more likely that the name was intended to serve an aetiological function in that Eleutherae was said to have been a mythical birthplace of Dionysos, or at least a town that was founded by the god.[36] Eleutherae stands over the fortified pass of Gyphtokastro, which leads to Mt. Cithaeron, especially sacred to Dionysos and the place where Pentheus was torn apart by the Bacchae. One of the myths associated with the foundation of the festival of Dionysos explains how a man named Pegasos, whom Pausanias associated with King Amphiktyon (c. 800 BCE),[37] brought a wooden image of Dionysos to Athens from Eleutherae in order to establish the god's worship. The Athenians refused to observe this new god, so Dionysos caused a disease to strike at the men's genitals. Once the god was duly worshipped, the terrible affliction cleared up and the Athenians paid homage to the god with phalloi.[38] Likewise, the procession of the *eisagoge* ("introduction") that saw the cult statue of Dionysos carried into the city from the Academy that lay outside of the city walls was a ritual embodiment of this aetiological myth that performed a representational connection between one of the innermost parts of the city of Athens and one of its most outlying, mountainous districts. Thus, as Dionysos was the god of wild places, mountainsides, and borderlands with mythological connections to Greece via Thebes, so Eleutherae stood on the border between Athenian and Theban territories. In this way, the site of the sanctuary and its *theatron* encapsulated the liminal spirit of Dionysos and connected its spectators to both the city it stood in and the wider – and wilder – countryside of Attica.

This name of the sanctuary therefore strongly indicates that the location was chosen for its associations with Dionysian liminality by means of its view. I wish to take this concept further and suggest that the view significantly enhanced a fundamental aspect of Dionysian cult practice, ingrained in the function of Greek drama – the presentation, performance, and acceptance of alterity and its related emotional effects. To explain this, we need to re-think what we mean when we describe ancient Greek drama as theatre in the "open air."

Re-thinking open-air theatre

In an essay on the Greek theatre, Roland Barthes summed up the experience of the open-air performance beautifully:

> We must not forget that the meaning of "open air" is its fragility. In the open-air the spectacle cannot be a habit, it is vulnerable, hence irreplaceable: the spectator's immersion in the complex polyphony of the open-air (shifting sun, rising wind, flying birds, noises of the city) restores to the drama the singularity of an event. The open-air cannot have the same image repertoire as the dark theater: the latter is one of evasion, the former of participation.[39]

Barthes' description highlights the sense of open-air theatre as a singular event, something that can never be repeated. This is applicable to all live performance, interior or exterior, as Philip Auslander has pointed out, coining the term "liveness" to describe a performance in its original social, political, and environmental context.[40] I remember staging a production of Aristophanes' *Clouds* in the ancient stadium at Delphi, nestled in the epic slopes of Mt. Parnassus. The surrounding environment could not be ignored but had to be incorporated into the visual schema of the show, just as Aristophanes did by having his Socrates beckon Strepsiades to look up into the sky above the Athenian Mt. Parnes to watch the chorus of clouds approach from the sky (323–324). Barthes is correct in describing the lack of control an artist can exert over a live performance in the open air, whether it was the sudden gusts of wind that ruffled my own show at Delphi or the kind of "complex polyphony" he describes. A show staged outside cannot be divorced from its open environment; the stadium at Delphi presented a view of the mountain side and it was not used in antiquity as a theatrical space, whereas the Hellenistic theatre at Delphi does command the most incredible view of the sky. Nevertheless, I will never forget the sense of spirituality in that ancient space, the beautiful fragility of that little show we presented there and the ways in which the textual references in the play took on a whole new relevance, especially when the chorus of clouds sing about looking down on Athens:

> The godly processions to sacred sites
> The splendid sacrifices that crown the land.
> Celebrations held throughout the year
> Then Sweet Dionysos comes in spring,
> And the resonant tones of pipes we hear
> As the joyous chorus dance and sing.
> Aristophanes, *Clouds*, 307–313

In *Clouds*, an old Athenian is invited by Socrates to look up and behold the alternate reality that the chorus of clouds will bring from the sky (323–330). The same thing happens in *Birds*, where another Athenian is invited to look up and envision an entire alternate fantastical polis called *Cloudcuckooland*, and in *Knights* an Athenian stares up into the sky and then mounts a giant beetle to fly up into the sky to seek an alternative to war. For Aristophanes Sky-space is frequently the realm of possibilities where the impossible can be contemplated and perhaps even come to pass.

The Greek terms for sky – *ouranus, olympus,* and *ether* – were interchangeable with the word for "heaven" or the place where the gods resided. Zeus was the sky-god whose name derived from the Indo-European *deus pitar* (sky father), from where his Latin name Jupiter is derived.[41] For the Greeks and many human cultures, the sky is the realm of the divine, gods are said to be "on high," "above," and "looking down on us," and the religions and spiritual practices of most cultures emphasize a vertical relationship to their respective deities. On a cognitive level, sky-space is where many people seek inspiration – we look up to disengage from the material world around us to seek alternate and imagined realities. Many of us embody this spatial cognitive relationship when we think deeply on a matter. There is also now clinical evidence that people perform better at memory retrieval when they look up or away into "nothing" space. This phenomenon has been termed LAN (Looking At Nothing). Here the researchers found that gaze direction was determined by the spatial relationship of the memory being retrieved.[42] Looking up into sky-space would then be indicative of retrieving the most abstract of memories or associations.

Most people identify the act of looking up with deep thought and contemplation. For example, when examining Baron-Cohen's "Reading the Mind on the Eyes Test," originally developed in 1997 and revised in 2001 to test for individual deviations in social sensitivity for adults with Asperger syndrome or high-functioning autism, most people report the upward-looking eyes as indicative of thinking, and this has been shown to be true of children as young as eight years old.[43] In studies involving people contemplating mathematical problems and number processing, it has been shown that there is a close relationship between verticality and quantity, in that smaller numbers will tend to be horizontally mapped (with gaze direction to match) while larger, more abstract numerical representations will be processed along with vertical eye movements.[44] It seems that when we need to engage with abstract concepts, we look up and form cognitive correlations between conceptual thought and "distal extrapersonal space" – the far-off landscape and the sky.

We find descriptions of this cognitive juncture among spatial distance, sky-space, and abstract thought in the metaphysical theories of Plato. For example, he depicts Socrates as standing for long periods gazing up at the sky deep in thought in the *Symposium* (175a-b and 220c), but this idea is expressed most explicitly in Plato's famous "Allegory of the Cave" in the *Republic*. In this story, people are compared to prisoners chained in place and forced to face in only one direction at a wall where only flickering shadows are projected. But if they could only turn their heads and look above they would see the light, and if one of their number broke loose and climbed up and out of the cave, he might contemplate the reality above. According to Plato, "the ascent and the contemplation of the things above is the soul's ascension to the intelligible region" (517b). For Plato, the contemplation of the realm above is where the "forms" dwell, the true elements of which everything below on earth is but an inferior copy. A person whose vision is focused upward and away from the "baseness" of ordinary human activity below will start to find the good, whereas the keen-sighted man who focuses his vision only on what is below will engage in evils such as gluttony and base pleasures (519b).

Plato is even more explicit in *Phaedo* (109b): humans living on earth are described as dwelling in "many hollows of various forms and sizes," and these are envisioned as actually being below the real surface of the earth, which is "situated in the pure sky." Plato goes on to compare men dwelling on earth as being akin to fish living in the sea: they can look up and faintly discern an upper world, but only when they leap up and break the surface of the water and see the air above can they begin to comprehend the "true sky, the true light and the true earth" (109e). In this vision of the world, the realm of human habitation is mired in brine and mud, while in the world above can be found the true beauty (110a). Plato's account of the upper and lower worlds in *Phaedo* creates an ethical map of human existence – it is only by searching for the truth high up in the sky that a mortal may begin to comprehend the good. The danger is that once these contemplative heights have been attained, the philosopher may never wish to return to the base earth; blinded by the light of the sky he may never be able to see clearly below. Even if the escaped prisoner of the cave returns to his fellow mortals, how can they believe what he would tell them of his spectacular journey? Yet tell them he must for according to Plato this is the duty of the philosopher – to take the knowledge gained in the celestial realm and use it to challenge the norms of the mortal world and to attempt to effect positive change.

Andrea Nightingale has shown how Plato sets his accounts of the contemplative skyward journey of the philosopher and his return back down to earth within the cultural context of Greek theoric religious festivals.[45] Theoric events involved a select group of ambassadors from a particular polis journeying as pilgrims to an important Panhellenic shrine, such as the Sanctuary of Apollo at Delphi, or Zeus at Olympia, or the sacred Island of Delos. There they would attend athletic events and performances, and watch and participate in processions where they would engage in what Jas Elsner has described as "ritual-centered visuality.[46] Nightingale's account of this type of "sacred viewing," influenced by the work of Elsner, Rutherford, and Jameson, can be set out in three phases:[47]

1   Journey – "the *theoros* departed from the social and ideological space of his city and entered a panhellenic space."
2   Sacred space – by entering a defined sacred space and viewing rituals and objects, there "the *theoroi* participated in a ritual event that transcended – and to some extent, challenged – the social political and ideological structures of any individual city."
3   Return – "By participating in a panhellenic event, he is confronted with difference and alterity and is himself altered by this experience. He thus returns home with a broader world-view and brings alterity to the city." In most cases, the returning *theoros* must give an account of what he has experienced to the citizens of his polis.[48]

Nightingale calls this Plato's "rhetoric of estrangement," and accounts such as the allegory of the cave serve to generate unsettling feelings of unfamiliarity, displacement, and alterity.

These three phases of *theoria* can be applied to the fifth-century Athenian theater. We know that a large portion (perhaps nearly one-quarter) of the audience were theoric guests coming from the many Allied states and that the Dionysia did involve a symbolic journey called the *eisagoge* when the statue of Dionysos was removed from its sanctuary and escorted back into the city as if it had come from elsewhere and had newly arrived.[49] Moreover, the participants in the festival also took part in a great procession, which provided another symbolic journey and involved the social reordering inherent in the ribald *komos* with the carrying of the phallus pole. Furthermore, the theatre was part of the sacred grounds of the Sanctuary of Dionysos Eleuthereus, named for the Attic community that was the farthest from Athens, on the border with Boeotia to the north. The brilliance of the idea behind the founding of the City Dionysia around 530 BCE was to provide a state-run *theoria* that promoted Athens as a cult center, radiating its authority to the various *demes* of Attica, near and far. Thus, to attend the Dionysia was certainly to go on a journey, both physically and imaginatively.

When an Athenian entered the sanctuary, what perceptual stimuli created the kind of otherworldly and alternate experience encountered by the *theoric* pilgrims who traveled to sanctuaries abroad? Of course, we can consider the performances as providing a form of aesthetic estrangement. For example, the dithyrambic choruses experimented with new musical styles and strove to be innovative, and tragedy placed masked mythological characters engaging in taboo activities before a live audience and presented music, dance, and the mimetic customs of women, foreigners, social outcasts, and slaves, what Froma Zeitlin termed "playing the Other."[50] Likewise, comedy engaged in ribald taboo breaking, political and social satire, and transgressive behaviors, and it explored fanciful, escapist, and dream-like fantasies, such as creating kingdoms in the clouds (*Birds*), fantastic journeys to the underworld (*Frogs*), and rides into the sky on massive insects (*Peace*). Finally, satyr dramas involved the tragic casts in the performance of sheer Dionysian anarchy and phallic ridiculousness. Performances at the Dionysia were transgressive, exciting, otherworldly, and strange, but as I will show, like Plato's theoric traveler, these kind of dissociative, absorbing, aesthetic experiences could induce deep emotions and feelings of empathy.

This sense of sacred space, where liminal transgressions could occur, extended beyond the confines of the theatre and sanctuary and encapsulated the entire sky, which was the most prominent feature in the visual field of the spectator, who by the very act of watching was also performing a sacred theoric duty at the Dionysia. Just as sky-space forms the realm of contemplative truth for Plato and is the place where the true forms reside, so in the Greek theatre the sky served a similar cognitive function by creating a sense of spatial dissociation that can contribute to the altering of mental states and promoting of feelings of spirituality and of the divine.

Two examples from Greek drama – one comic, the other tragic – illustrate the cognitive properties of sky-space to impact and alter human states of consciousness. In Aristophanes' *Clouds*, Socrates famously enters on the *mechane* (the stage crane) suspended over the performance space on a drying rack as if he were a large prune drying out in the sun. He explains to the bemused old man, Strepsiades, that he is trying to remove himself from the moisture of the earth and elevate his mind:

> In order that I may make exact discoveries of the highest nature!
> Thus, my mind is suspended to create only elevated notions.
> The grains of these thoughts then merge with the similar
> atmosphere of thin air! If I had remained earthbound
> and attempted to scrutinize the heights, I would have found
> nothing; for the earth forces the creative juices to be drawn
> to its core, just as it is with Watercress!
>
> Aristophanes, *Clouds*, 228–234

Socrates' comic description of his ridiculous entrance is reminiscent of a vivid description of human-embodied cognition in Plato's *Timaeus* (90a):

> We declare that God has given to each of us, as his daemon, that kind of soul which is housed in the top of our body and which raises us – seeing that we are not an earthly but a heavenly plant up from earth towards our kindred in the heaven. And herein we speak most truly; for it is by suspending our head and root from that region whence the substance of our soul first came that the Divine Power Keeps upright our whole body.

By suspending himself above the earthly realm on a drying rack, Aristophanes' Socrates hopes to avoid the fate of the Watercress plant, whose roots are mired in moisture, and fully elevate his thinking into the realm of the sky, where he can "speculate on the sun." This is similar to another fantastical Aristophanic *mechane* scene in *Peace*. Here, the comic hero Trygaeus embarks on a personal *theoria*, up into the sky on the back of a huge dung beetle to consult with Zeus and try to stop the Peloponnesian War (82–153).

My tragic example is from Euripides' *Baccahe* (1270). Dionysos has driven the women of Thebes into a collective ecstatic altered state in the mountains, and they have torn apart the defiant young king, Pentheus, thinking him a mountain lion. The mother of Pentheus, Agave, returns to Thebes clutching her son's severed head. She is told by Cadmus to look up at the sky and then back down at the head (probably his mask), so she will come to her senses and see clearly that it is her son's head and not that of a lion.

Cadmus: First let your eyes look at the sky. Up here.
Agave: I'm looking, why do you suggest I look at this?
Cadmus: Is it the same? Or do you think it changes?
Agave: It's brighter than before, a new glow comes through it.
Cadmus: And that fluttering sensation, still have it in your soul?
Agave: I don't know what you mean. But I am somehow coming back into my
            mind, I'm moving away from the old thoughts.

Euripides, *Bacchae*, 1264–1270, tr. P. Woodruff

When Agave looks up into the sky, her state of mind changes, and she recognizes the reality of what she holds in her hands. This engagement with the ambient extrapersonal sky-space has altered her mental state – from an ecstatic wild devotee of Dionysos to a mortal mother looking upon her mutilated son.[51] This cognitive contact with sky-space has changed her, the open air of the physical theatre space being conflated with the sky over mythical Thebes and the state of mind of a tragic character.

The four realms of theatrical three-dimensional space

These passages indicate that the Greeks knew the efficacy of sky-space as the realm of deep contemplation and alternate views. Our understanding of the cognitive effects of this spatial realm was significantly enhanced by Fred Previc, a senior researcher with the United States Air Force Research Laboratory in San Antonio, Texas. Previc headed up the Spatial Disorientation Countermeasures Task Group, which was given the task of reducing the amount of aircraft crashes caused by pilot disorientation. This is when a pilot becomes completely cognitively disengaged from his or her position in space and then loses control of the aircraft because of the loss of the sense of how to orientate properly in relation to the ground – this is commonly known as "losing the horizon" or "pilot disorientation."

Previc's work was intended to help understand why these kinds of crashes were happening and to help devise instrumentation and pilot training to counter them. What he developed was a revolutionary way to examine the relationship of the body in space. Previc postulated that spatial disorientation was a cognitive state connected to the body's relationship to the space around it. His model assumed that the environment as it is processed by human perceptual systems is actually a creation of those perceptual systems based on the human body (Fig. 2.2). He then developed an embodied system of four realms of three-dimensional space to more fully understand this phenomenon. These realms break down as follows:[52]

- ***Peripersonal space*** – the field in which cognitive processing of items within our physical reach takes place, such as holding and manipulating objects
- ***Focal extrapersonal space*** – the area immediately beyond our physical reach, where eye movements are used for visual scanning and object recognition
- ***Action extrapersonal space*** – assists humans to orientate themselves within topographically defined space. This is the way in which we use landmarks to navigate and place ourselves within a wider spatial context than in the more immediate peripersonal or focal exptrapersonal fields.
- ***Ambient extrapersonal space*** – the spatial field furthest away from our own bodies – the sky, distant mountains, the horizon – the place where Previc's pilots were becoming disorientated and crashing their planes

Theatre happens in each of Previc's four realms: (1) the exchange of stage properties or gestures in peripersonal space; (2) the mask operating in focal extrapersonal space; (3) the significant sanctuaries and cult sites within the optic array of action in extrapersonal space; and what we will be concentrating on here, (4) the place of abstraction, divinity, and the role of dopamine in ambient extrapersonal space, which in the ancient theatre was the sky. Ambient extrapersonal cognition

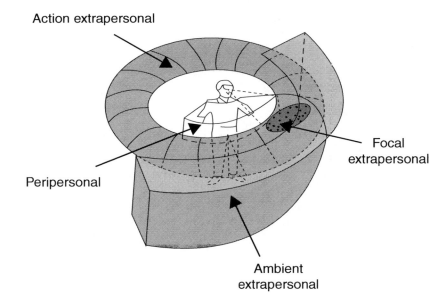

Figure 2.2 Previc's four realms of three-dimensional space.

is described as a person's relationship to the most distant abstract spatial realm where one feels most disconnected from his or her own body. This is where we can experience "out-of-body" feelings, disorientation, and dizziness; it is the cognitive space where Previc's pilots were becoming discombobulated and crashing their flight simulators. In other tests involving pilots who were discombobulated in simulated ambient extrapersonal space by centrifuges, many pilots also reported having out-of-body experiences and feeling a sense of "the divine."[53]

According to Previc, extrapersonal space promotes a sense of the divine and "dreams and hallucinations represent the triumph of the extrapersonal systems over the body-oriented or peripersonal systems."[54] He notes how spatial language such as "on high" and "exalted" are frequently used to describe the divine, with language of "highness" being generally held as positive, such as in Euripides' *Bacchae* when Dionysos says "the fame you will find will rise up to heaven" (972). The opposite of this vertical axis is usually negative, such as in Aeschylus's *Eumenides* where the chthonic Furies say "we are called curses under the earth" (417). A vertical spatial dichotomy is commonly found in Greek literature where Olympian gods are placed above while darker, primordial deities such as the Furies are associated with the underworld. We can place this basic spatially organized belief system in the context of Previc's work:

A review of the neuroanatomy and neurochemistry of dreaming, hallucinations, and religious beliefs, practices and experiences in normal humans indicates that there may be a common neural substrate of all behavioral phenomena that reflect a predominance of extrapersonal brain system activity and a reduction of bodily (self-oriented) activity.[55]

Figure 2.3 The Theatre of Dionysos in Athens (photo by the author).

For the spectators seated in the Dionysian *theatron*, the horizon and sky dominated their ambient extrapersonal visual field, and this is one of the most significance differences between the modern stage and the ancient Greek stage. The sense of religiosity and spiritually that pervades Greek drama was enhanced not only by the *orchestra*'s relationship to the sanctuary below, the festival atmosphere, and the visual relationship of the audience to places of cult significance, but also, and perhaps most importantly in terms of neural chemistry, by the cognitive predominance of the extrapersonal space of the sky.

Dionysos and dopamine

Today, most theatres place us in a darkened room, illuminate the playing area with artificial light, and deny us any relationship with the external environment. Occasionally and accidently, the closed relationship between the world of the play in the theatre and the world outside is ruptured by a passing siren or some other such loud noise. We tend to find this jarring – an interruption – and it can take us out of the experience of the play for a while. When we are *in* the theatre we usually don't want the outside world to permeate its walls. The opposite was true in the *theatron* at the Sanctuary of Dionysos Eleuthereus: there was no division between inside and outside; the world and the play were deliberately visually, aurally, and olfactorily blurred, as was time and place, myth and reality, the world of humans and the realm of the divine.

Previc also showed that ambient extrapersonal space plays an important role in the neurochemistry of religious activity. Concepts of space perception have shifted radically over the past 20 years, and the neural processing of space is seen less as a "mapping" of external space and more as "the locus of integration between perception, action and awareness."[56] Neural coding of an object in external space is related to the neural functions of grasping and other bodily movements. Our mental concept of external space is dependent on several interdependent elements that process perception and "error-correct" by deploying action inferences. Thus, the space nearest to our bodies, peripersonal space, is where grasping and touching occurs, whereas extrapersonal space – distant and sky-space – is physically out of reach and therefore the most abstract. Contemplating extrapersonal space seems to involve "the tendency of projecting the self into mental dimensions that transcend sensorimotor contingencies."[57] This has been supported by transcranial magnetic stimulation studies that create temporary virtual lesions in the areas of the brain that have been associated with the processing of extrapersonal space, including the inferior parietal lobe and the dorsolateral prefrontal cortex.[58] The virtual lesions affected how participants performed when expressing spiritual or religious concepts such as "divine," "spirit," "supernatural," and "soul." Of course, the conflation of neuroscience with religious experience is a very slippery topic, but studies like this indicate that the close connection between the mind's neural systems for the processing of extrapersonal space and at least the conceptualization of spirituality are connected with that spatial field.

There are multiple neural networks and brain areas connected to the processing of external space and complex electrochemical interactions that contribute to the operation of these functions. One important neural transmitter that has been identified as part of this complicated process is dopamine, which has a functional relationship to the processing of movement and prediction formulation.[59] The ventral dopaminergic pathways, which transmit dopamine between brain areas, have also been shown to be biased toward ambient extrapersonal, distant, and especially upper spatial fields, associated with abstract thought, hallucinations, and dreams.[60] Dopamine also helps transmit communication between nerve cells and to muscle cells that are responsible for movement, motivation, appetite, and attention. For example, attention deficit disorder has been linked to low levels of dopamine in the prefrontal cortex, and the loss of motor function with Parkinson's disease may also be linked to depleted levels of dopamine. Working memory is also dependent on dopamine, which is released in the prefrontal cortex to provide a kind of "gating" system to regulate the contextual processing of memory, an essential component in the top-down portion of the predictive generative model.[61]

In a 2005 study, Previc and his team posited a strong link between higher-order cognition or abstract thought and perceptual motor actions in three-dimensional space.[62] Previc's four spatial realms are perceived by humans on a vertical axis, and this is due to the way in which we visually process our environment. Grasping and reaching in peripersonal space mainly occurs in the lower visual field, below eye level (we look down), but what Previc describes as "the slope of the visual world away from the person" or extrapersonal distant space is located in the upper visual field (we look up). Thus, ambient extrapersonal space – sky-space – is where we locate deep thought and contemplation. We attempt to engage with ambient extrapersonal space whenever we need to think deeply about something. Notice how people's gaze direction tends to glance up when having a specific thought, even when exterior sky-space is not in the current visual field. You can try this simple experiment to illustrate what I mean: Ask somebody to tell you their favorite food. Then watch their face as you ask the next question – "when did you last have that?" Most people will spontaneously glance upwards, as they think a little more deeply, their gaze direction seeks out extrapersonal ambient space – the place of contemplation and of memory.

This autonomic eye saccade links higher-order cognition to extrapersonal space and working memory. Humans possess the ability to create abstract thoughts, and this kind of "distant orientation" is essential to mental visualization and the recall of memories that can transcend both spatial and temporal frames of reference. In physiological terms, the area of the brain responsible for regulating these kinds of higher-order eye movements and certain functions of working memory is the dorsal lateral prefrontal cortex (DLPFC), which is situated in front of the frontal eye fields (FEF). The DLPFC is a dopamine-rich environment, and the dopamine located here is the main neurotransmitter of the focal-extrapersonal space pathways used for controlling the movements of the eyes when scanning distant space.[63] The act of looking up and into LAN space not only creates the sense of perceptual disembodiment but also releases dopamine to facilitate abstract

thought, memory retrieval, and advance planning. This is also attested to in animal studies where increased dopamine levels in rats led to a wider range of forag-
ing, and monkeys explored beyond their ordinary territorial boundaries.[64] This
provides us with a neurobiological reason why the expansive views of classical
Greek theatres situated extrapersonal sky-space as the most prominent element in
the audience's optic array.[65]

Dopamine is instrumental in both religious thought and extra-human or "out-of-
body" behaviors that are also often associated with religious experience including
dreams, hallucinations, schizophrenia, and seizures. One study led by Krummenacher in 2002 showed that when a dopamine precursor l-dopa, which acts like
dopamine, was given to people who described themselves as skeptical, their perceptual sensitivity decreased to the same level as that of paranormal believers
(this category includes people who describe themselves as religious).[66] This study
posited that paranormal thinkers had higher than normal levels of dopamine, and
this produces a distortion of extrapersonal sensory inputs promoting superstitious
beliefs and a characteristic underestimation of the role of chance.[67] The panoramic
views of the sky available to the audience in the Theatre of Dionysos offered
something that is denied to most modern theatregoers today: it allowed them to
both watch the play before them and to deeply contemplate what they experienced
in a space where thoughts were not constrained by scenery or walls, where spirituality could flourish, and minds could be changed.

The dopamine-inducing environment of the ancient theatre and the way in
which it promoted alternate modes of thought and contemplation also has a strong
neural link with the mechanisms of predictive processing. Wolfram Schulz has
recently written:

> Having a neuronal correlate for a positive reward prediction error in our
> brain [dopamine] may explain why we are striving for ever-greater rewards, a
> behavior that is surely helpful for surviving competition in evolution, but also
> generates frustrations and inequalities that endanger individual well-being
> and the social fabric.[68]

Dopamine is not the only neurotransmitter that has been linked with predictive
processing and the means by which we generate uncertainty (error correction)
about states. Friston also names serotonin, acetylcholine, and noradrenalin as
modulatory neurotransmitters that are induced by randomness in the world.[69] Yet
he also posits that dopamine plays a central role because prediction errors cannot
be resolved alone by referencing only top-down precise prior memories, so we
also need active inference. For this we need action, either imagined or carried
out, for which dopamine is essential, as it facilitates neural transmission between
the cognitive systems that involve movement, planning, and abstract thought and
also encodes corrected prediction errors by reward. The reward of experiencing ancient theatre was in the confrontation of transgression and the excitement
of surprisal, the provoking of emotions, and the eliciting of empathy. Schulz's
description of the positives and negatives of dopamine and its connections to both

prediction and addiction sound a lot like the thrills, joys, tensions, and ambiguities found in Greek drama and the personality traits of the god Dionysos himself.

Previc solved the pilot disorientation problem by moving key instruments from the peripersonal panel up onto the cockpit windshield by means of projection. Thus, the pilot did not become spatially confused by constantly switching between the ambient extrapersonal sky space and the peripersonal space within reach. On the other hand, the expansive open-air venue of the *theatron* at the Sanctuary of Dionysos Eleuthereus promoted a sense of disorientation, possibility, and spirituality, although not as extreme as in the case of Previc's pilots. The spectators were placed within a gentler dopamine-promoting environment that fostered mind-expanding thoughts. This was a key ingredient in the Greek theatre's ability to take its audience on journeys of alterity and to open their minds to different perspectives – the environmental starting point for a mimetic journey of alterity, dissociation, absorption, and empathy. How then, in this large open-air space, were the Athenian dramatists able to create levels of deeply felt emotional intensity emanating from the actions of the characters depicted in their works? This was in large part due to the mask and how it affected the reception of movement, which, rather than distancing the audience or simply disguising the performer, was actually able to seemingly change its expressions and serve as the emotional face of the theatre. In *Poetics*, Aristotle writes that the soul-enthralling qualities of *opsis* cannot be ordered by the poet but are the preserve of the *skeuopoiou*, the mask-maker (1450b20). In the next chapter we will explore the uncanny attributes of the Greek dramatic mask.

Notes

1 On this aspect of *opsis* in *Poetics*, see Calame 2005: 106–107; Taplin 1977: 477–479; Halliwell 1986: 337–343; Bassi 2005: 251–270.
2 Gibson 1977: 67–82.
3 Rehm 2002: 14–15.
4 Rangacharya 1998: 73–81.
5 Schechner 2004: 29.
6 Rangacharya 1998: 75.
7 Friston et al. 2012.
8 Anaximenes 1428b5–10.
9 Moss 2012: 6–9.
10 Moss 2012: 49–50.
11 Moss 2012: 53–54.
12 Schulz 1998.
13 Friston et al. 2012.
14 FitzGerald et al. 2015.
15 Pasquereau and Turner 2015.
16 Artaud 1958: 37.
17 Csapo 2007: 92–121; Roselli 2011; Paga 2007: 351–384; Bosher 2006: 151–160, tables 6.1, 6.2, and 6.3.
18 Anti 1947; Gebhard 1974: 428–440; Wiles 1997.
19 Ober 2008: 199–200.
20 Goldhill 1987: 58–76.
21 Csapo 2007: 92–121.

22 Csapo 2007: 92–121.
23 Papastamati-von Moock 2014; Meineck 2012: 20–23; Goette 2007: 116–121; Moretti 2001. Dawson has placed the number much lower at 3,700. Dawson 1997: 1–14.
24 For example, Doerries 2015: 38, who writes that "nearly a third of the citizens in Athens" gathered to watch the plays. If we assume the generally accepted figure of 40,000–60,000 in the mid-fifth century, this would be at least 15,000–17,000 spectators, far too large for the fifth-century evidence. On the size of the Athenian population, see Roselli 2011: 13 and 206 n.45 and n.46.
25 Pickard-Cambridge 1956: 141 and n.2.V.
26 On the reforms of Cleisthenes, see Pritchard 2004: 208–228. For evidence of a "Periclean" renovation, see Papastamati-von Moock 2014.
27 Carter 2004; Tzanetou 2012.
28 I thank Kurt Raaflaub for pointing this out to me.
29 Bosher 2006; Paga 2010: 351–384.
30 For a full account of these entries and how they have been misunderstood, see Meineck (2012).
31 Noy 2002: 176–185. For excellent discussions of the use of the *eisodoi*, see Taplin 1983: 155–183; and Rehm 2002: 20–25.
32 Connor argues for a date of around 500 BCE (Connor 1990); however, Cartledge, who finds much to admire in Connor's arguments, places the date of the foundation of the City Dionysia around 530 (Cartledge 1997: 23–24).
33 Kowalzig 2007: 24–30.
34 Kowalzig 2007: 24–32. See also Sourvinou-Inwood 2003: 22–25, who described the perceptual frame of Greek drama as "zooming" between the mythic past and contemporary religious practices.
35 Connor 1990.
36 Diodorus 3.66.1 and 4.2.6.
37 Pausanias 1.2.5.
38 The story is found in the Scholion to Aristophanes' *Acharnians* 243.
39 Barthes 1985: 79.
40 Auslander 2008: 52.
41 Burkert 1985: 125–130.
42 Scholz et al. 2016.
43 Baron-Cohen et al. 2001: 42, 241–252; Previc et al. 2005.
44 Winter and Matlock 2013.
45 Nightingale 2004: 40–71.
46 Elsner 2007:1–28.
47 Rutherford 1998; Jameson 1999.
48 Nightingale 2004: 45–48.
49 Sourvinou-Inwood 2003: 89–100.
50 Zeitlin 1996.
51 When Pentheus emerges dressed in women's clothing ready to go to the mountain to spy on the women, his altered state is also expressed in ambient extrapersonal terms when he exclaims "I can see two suns in the sky!" (Euripides, *Bacchae* 920).
52 Previc 1998: 123.
53 Whinnery and Whinnery 1990: 764–776.
54 Previc 2006: 510.
55 Previc 2006: 518.
56 Vallar and Maravita 2009
57 Crescentini et al. 2014: 2.
58 Crescentini et al. 2014: 1–15.
59 Collins 2016: 3–5.
60 Previc 2006: 500–539.

61 D'Ardenne et al. 2012.
62 Previc, Declerck and de Brabander 2005: 7–24.
63 Previc 1998: 123–164.
64 Previc 2006.
65 Previc 2009: 38–40.
66 Krummenacher et al. 2010: 1670–1681.
67 Blackmore and Trościanko 1985: 459–468; Brugger and Graves 1997: 55–57.
68 Schultz 2016: 23.
69 Friston 2009: 293–301.

Bibliography

Anti, C. 1947. *Teatri greci arcaici da Minosse a Pericle*. Padova, Italy: Le Tre Venezie.
Artaud, A. 1958. *The theater and its double*. New York: Grove Press.
Auslander, P. 2008. *Liveness: Performance in a mediatized culture*. London/New York: Routledge.
Baron-Cohen, S., Wheelwright, S. and Hill, I. 2001. "The 'reading the mind in the eyes' test revised version: A study with normal adults and adults with asperger syndrome or high-functioning autism" in *Journal of Child Psychology and Psychiatry* 42: 241–252.
Barthes, R. 1985. *The responsibility of forms*. Howard, R. (Tr.), Berkeley, CA: UC Press.
Bassi, K. 2005. "Visuality and temporality: Reading the tragic script" in Pedrick, V. and Oberhelman, S.M. (Eds.), *The soul of tragedy: Essays on Athenian drama*. Chicago: University of Chicago Press: 251–270.
Blackmore, S. and Trościanko, T. 1985. "Belief in the paranormal: Probability judgements, illusory control, and the 'chance baseline shift'" in *British Journal of Psychology* 76.4: 459–468.
Bosher, G.K. 2006. *Theatre on the periphery: A social and political history of theater in early Sicily*, PhD Thesis, The University of Michigan.
Brugger, P. and Graves, R.E. 1997. "Right hemispatial inattention and magical ideation" in *European Archives of Psychiatry and Clinical Neuroscience* 247.1: 55–57.
Burkert, W. 1985. *Greek religion*. Cambridge, MA: Harvard University Press.
Calame, C. 2005. *Masks of authority: Fiction and pragmatics in Ancient Greek poetics*. Ithaca, NY: Cornell University Press.
Carter, D. 2004. "Was Attic tragedy democratic?" in *Polis: The Journal of the Society for Greek Political Thought* 21.1–2: 1–25.
Cartledge, P. 1997. "'Deep plays': Theatre as process in Greek civic life" in Easterling, P.E. (Ed.), *The Cambridge companion to Greek tragedy*. Cambridge: Cambridge University Press: 3–35.
Collins, A.G.E. and Frank, M.J. 2016. "Surprise! Dopamine signals mix action, value and error" in *Nature Neuroscience* 19.1: 3–5.
Connor, W.R. 1990. "City dionysia and athenian democracy" in Connor, W.R., Hansen, M.H., Raaflaub, K.A. and Strauss, B.S. (Eds.), *Aspects of Athenian democracy*. Copenhagen Museum Tusculanum Press: 27–33.
Crescentini, C., Aglioti, S.M., Fabbro, F. and Urgesi, C. 2014. "Virtual lesions of the inferior parietal cortex induce fast changes of implicit religiousness/spirituality" in *Cortex* 54: 1–15.
Csapo, E. 2007. "The men who built the theaters: *Theatropolai, Theatronai*, and *Arkhitektones*" in Wilson, P. (Ed.), *The Greek theatre and festivals: Documentary studies*. Oxford: Oxford University Press: 97–121.

D'Ardenne, K., Eshel, N., Luka, J., Lenartowicz, A., Nystrom, L.E. and Cohen, J.D. 2012. "Role of prefrontal cortex and the midbrain dopamine system in working memory updating" in *Proceedings of the National Academy of Sciences* 109.49:19900–19909.

Dawson, S.M. 1997. "The theatrical audience in fifth-century Athens: numbers and status" in *Prudentia* 29.1: 1–14.

Doerries, B. 2015. *The theater of war: What ancient Greek tragedies can teach us today.* New York: Knopf.

Elsner, J. 2007. *Roman eyes: Visuality & subjectivity in art & text.* Princeton, NJ: Princeton University Press.

FitzGerald, T.H., Dolan, R.J. and Friston, K.J. 2014. "Dopamine, reward learning, and active inference" in *Frontiers in Computational Neuroscience* 136: 1–16.

Friston, K.J. 2009. "The free-energy principle: A rough guide to the brain?" in *Trends in Cognitive Sciences* 13.7: 293–301.

Friston, K.J., Shiner, T., FitzGerald, T.H., Galea, J.M., Adams, R., Brown H., et al. 2012. "Dopamine, affordance and active inference" in *PLoS Computational Biology* 8.1: 1–20.

Gebhard, E. 1974. "The form of the orchestra in the early Greek theater" in *Hesperia* 43.4: 428–440.

Gibson, J.J. 1977. "The theory of affordances" in Shaw, R. and Bransford, J. (Ed.), *Perceiving, acting, and knowing: Toward an ecological psychology.* Hillsdale, NJ: Erlbaum: 67–82.

Goette, H.R. 2007. "Archaeological appendix" in Wilson, P. (Ed.), *The Greek theatre and festivals: Documentary studies.* Oxford: Oxford University Press: 116–121.

Goldhill, S. 1987. "The great Dionysia and civic ideology" in *The Journal of Hellenic Studies* cvii: 58–76.

Halliwell, S. 1986. *Aristotle's poetics.* Chicago: University of Chicago Press.

Jameson, M.H. 1999. "The spectacular and the obscure in Athenian religion" in Goldhill, S. and Osborne, R. (Eds.), *Performance culture and Athenian democracy.* Cambridge: Cambridge University Press: 321–340.

Kowalzig, B. 2007. *Singing for the gods: Performances of myth and ritual in Archaic and classical Greece.* Oxford: Oxford University Press.

Krummenacher, P., Mohr, C., Haker, H. and Brugger, P. 2010. "Dopamine, paranormal belief, and the detection of meaningful stimuli" in *Journal of Cognitive Neuroscience* 22.8: 1670–1681.

Meineck, P. 2012. "The embodied space: Performance and visual cognition at the fifth century Athenian theatre" in *New England Classical Journal* 39: 3–46.

Moretti, J. 2001. *Théâtre et société dans la Grèce antique: Une archéologie des pratiques thèâtrales.* Paris: LGF/Le Livre de Poche.

Moss, J. 2012. *Aristotle on the apparent good: Perception, phantasia, thought, and desire.* Oxford: Oxford University Press.

Nightingale, A.W. 2004. *Spectacles of truth in classical Greek philosophy: Theoria in its cultural context.* Cambridge: Cambridge University Press.

Noy, K. 2002. "Creating a movement space: The passageway in Noh and Greek theatres" in *New Theatre Quarterly* 18.2: 176–185.

Ober, J. 2008. *Democracy and knowledge: Innovation and learning in classical Athens.* Princeton, NJ: Princeton University Press: 199–200.

Paga, J. 2007. "Mapping politics: An investigation of Deme theatres in the fifth and fourth centuries B.C.E" in *Hesperia* 79: 351–384.

Paga, J. 2010. "Deme Theaters in Attica and the Trittys System" in *Hesperia* 79.3: 351–384.

Papastamati-von Moock, C. 2014. "The theatre of dionysus eleuthereus in Athens: New data and observations on its 'Lycurgan' Phase" in Csapo, E., Goette, H.R., Green, J.R. and Wilson, P. (Eds.), *Greek theatre in the fourth century BC*. Berlin: De Gruyter: 15–76.

Pasquereau, B. and Turner, R.S. 2015. "Dopamine neurons encode errors in predicting movement trigger occurrence" in *Journal of Neurophysiology* 113.4: 1110–1123.

Pickard-Cambridge, A.W. 1956. *The theatre of Dionysus in Athens*. Oxford: Clarendon Press.

Previc, F. 1998. "The neuropsychology of 3-D space" in *Psychological Bulletin* 124.2: 123–164.

Previc, F. 2006. "The role of the extrapersonal brain systems in religious activity" in *Consciousness and Cognition* 15: 500–539.

Previc, F. 2009. *The dopaminergic mind in human evolution and history*. Cambridge: Cambridge University Press.

Previc, F., Declerck, C. and de Brabander, B. 2005. "Why your 'head is in the clouds'during thinking: The relationship between cognition and upper space" in *Acta Psychologica* 118: 7–24.

Pritchard, D. 2004. "Kleisthenes, participation, and the dithyrambic contests of late archaic and classical Athens" in *Phoenix* 58.3: 208–228.

Rangacharya, A. 1998. *Introduction to Bharata's Nāt□ya-śāstra: Adya Rangacharya*. London: Munshiram Manoharlal Publishers Pvt. Limited: 73–81.

Rehm, R. 2002. *The play of space: Spatial transformation in Greek tragedy*. Princeton, NJ: Princeton University Press.

Roselli, D. 2011. *Theater of the people: Spectators and society in ancient Athens*. Austin, TX: University of Texas Press.

Rutherford, I. 1998. "Theoria as theatre: The pilgrimage theme in Greek drama" *Proceedings of the Leeds International Latin Seminar* 10: 131–156.

Schechner, R. 2004. *Performance theory*. London/New York: Routledge.

Scholz, A., Mehlhorn, K. and Krems, J.F. 2016. "Listen up, eye movements play a role in verbal memory retrieval" in *Psychological Research* 80.1: 149–158.

Schulz, W. 1998. "Predictive reward signal of dopamine neurons" in *Journal of Neurophysiology* 80: 1–27.

Schultz, W. 2016. "Dopamine reward prediction-error signalling: A two-component response" in *Nature Reviews Neuroscience* 17.3: 183–194.

Sourvinou-Inwood, C. 2003. *Tragedy and Athenian religion*. Lanham, MD: Rowman & Littlefield.

Taplin, O.P. 1977. *The stagecraft of Aeschylus: The dramatic use of exits and entrances in Greek tragedy*. Oxford: Oxford University Press.

Taplin, O.P. 1983. "Sophocles in his theatre" in de Romilly, J. (Ed.), *Sophocle, Sept exposés suivis de discussions*. Geneva: Foundation Hardt 29: 155–183.

Tzanetou, A. 2012. *City of suppliants: Tragedy and the Athenian empire*. Austin, TX: University of Texas Press.

Vallar, G. and Maravita, A. 2009. "Personal and extrapersonal spatial perception" in Berntson, G.G. and Cacioppo, J.T. (Eds.), *Handbook of neuroscience for the behavioral sciences*. Hoboken, NJ: Wiley: 322–336.

Whinnery, J.E. and Whinnery, A.M. 1990. "Acceleration-induced loss of consciousness: A review of 500 episodes" in *Archives of Neurology* 47.7: 764–776.

Wiles, D. 1997. *Tragedy in Athens: Performance space and theatrical meaning*. Cambridge: Cambridge University Press.

Winter, B. and Matlock, T. 2013. "More is up . . . and right: Random number generation along two axes" in Salter, W. J. (Ed.), *Proceedings of the 35th annual conference of the Cognitive Science Society*. Austin, TX: Cognitive Science Society: 3789–3974.

Zeitlin, F. 1996. *Playing the other: Gender and society in classical Greek literature*. Chicago: University of Chicago Press.

3   *Ethos*

The character of *catharsis*

In *Poetics* Aristotle has it that because tragedy represents action performed by living people, the performers must have certain qualities of character (*ethos*) and intentions (*dianoia*). In this chapter I re-evaluate Aristotle's concept of *ethos* as it relates to the portrayal of a character by a masked actor. I turn to *dianoia* in Chapter Four and relate intentionality and empathy to the reception of movement. The questions I seek to answer here are why did the actors of Greek drama wear masks, what was their function, and how did their fixed surfaces project the kind of emotionality that Aristotle considered necessary for *catharsis* – emotional purgation or understanding? The word *character*, first attested in 14th-century English, is derived from the Greek *karakter*, which means to engrave, inscribe, or make a mark. The term implies something fixed or unchanging, as if one's character is essential. This suggests that the face of the mask was also invariable. I propose that this was not at all the case: the Greek word for mask (and face) was *prósōpon*, which means something like "before the gaze," and this reflects the mask's capabilities in acting as an effective material anchor for the projection of a variety of powerful emotional states.

The issue of character in Greek drama has been a topic of debate for many years now, and several scholars have discussed the use of the mask in this context.[1] Gould felt that masking denied the audience what he called "the flickering procession of ambiguous clues to inaccessible privacy,"[2] whereas Marshall rightly proposed that minimalist masks can actually convey a good deal about the psychological aspects of a character.[3] Wiles commented that character is an externalized event in Greek masked drama: the actor puts on a face and there is no other character apart from that face, and when the actor changes the mask, the character ceases to exist.[4] More recently, Seidensticker has claimed that the mask "deprives the actor of the almost infinite expressiveness of the human face" and that in the large open-air theatre, subtlety, such an important aspect of characterization in modern drama, was not possible.[5] I hope to show that this was not at all the case. Seidensticker does make the important point that while the production conditions of the Greek theatre did not allow "a detail loving realism," the necessary concentration on essentials did not mean that individuality of character was dispensed with. In this chapter, I hope to show that the abstract and stylized nature of the mask and the way it was used in performance actually significantly enhanced the ambiguity, nuance, and emotionality of the characters it represented.

There is a major stumbling block in seeking to understand the ancient Greek mask: no actual theatre mask dating from the fifth century BCE has survived. There is also a dearth of evidence both literary and material for Athenian masks from this time – the period of Aeschylus, Sophocles, Euripides, and Aristophanes. What little we do have is found on non-theatrical art forms, such as vase painting and sculpture, and a very few scant references to masks in texts.[6] Most modern popular notions of Greek masks, with their stony faces, gaping eyes, "megaphone" mouths, and elongated headdresses, come from the Hellenistic or Roman theatre, which used a very different theatrical aesthetic (Figure 3.1). This has led to a general lack of comprehension about how the mask worked in performance, as well as a plethora of misunderstandings and popular myths about masks that are still prevalent today.

These misunderstandings include that the mask was merely a prop or part of the costume[7] and that it disguised the performer and facilitated easy doubling of parts, but the Greek mask was far more about projection than disguise, and doubling of parts was a result of the mask, not a means to facilitate it. It is also not necessary for the cross-gendering of roles, after all this was also a feature of the unmasked Elizabethan theatre where young male actors would take on parts such as Lady Macbeth. Some scholars have cited its advantages in being able to help the audience see the actors in large open-air theatre spaces,[8] but this is incorrect: the fifth-century stage was open air and larger than most theatres today but far smaller than the vast Hellenistic stone theatres that did employ large masks. The evidence also indicates that the classical mask was no larger than the human head; it was the properties of the mask that created engagement, not its size, as the director Peter Hall has written of his masked productions in ancient theatres: "a mask of human scale is perfectly visible in Epidauros before 10,000 people."[9] One of the most egregious mask myths is that the actor's voice was amplified via a megaphone within the mask itself.[10] There is absolutely no evidence for this whatsoever, and my own work with reconstructions of fifth-century masks at several ancient theatres (see Figures 3.4 and 3.5) has shown that as long as the mask fits well and the actor faces out front when speaking, there are no audibility issues.[11] Speakers did not use megaphones in the Pnyx assembly, a comparably sized space, so why would they have been needed in the theatre?

Perhaps most of all it has been argued that the mask was a blank canvas, and its neutrality a major factor in the production, of what John Jones described as a "distancing effect" for the spectators.[12] Likewise, Stephen Halliwell concluded that its fixed features were unable to form different emotional aspects, and it conveyed a sense of "heroic dignity even in the midst of destructive sufferings."[13] Claude Calame described the mask as having a "mediating quality," believing it was a facility of its "blank" surface, and envisioned the mask punctured by the two eyeholes and mouth, which enabled the actor to be revealed.[14] Yet the material evidence we do have for fifth-century masks on around 20 vase paintings and relief sculptures shows that they had filled eyes with

Figure 3.1 Detail from the Pronomos vase showing an actor wearing a Herakles costume and Papposilenos, performers, faces of two chorus members, and a female mask. Attic red-figured volute krater by the Pronomos painter, 425–375 BCE. Courtesy of Museo Archeologico Nazionale, Naples (NM 81673).

prominent sclerae and only small pupils for eye holes. Being directly gazed on by a mask facing out toward the audience, which was necessary for audibility, would have been exceptionally compelling and would have increased spectator attentionality.[15]

So, "why mask?" In addition to the mask being an active symbol of drama's cultic associations, it provided a highly effective material anchor for the audience's own projection of emotion, which I will discuss in a later section, and was capable of displaying a plethora of changing affective states. This meant that it was one of the most powerful communicators of emotions of the ancient stage and a fundamental element in the development and successful reception of Greek drama in antiquity. The mask was so fundamental that no actor ever performed barefaced, Aristotle hardly mentioned it,[16] and it quickly became the very symbol of the theatre itself.

Mask as theatrical frame

I perform a simple experiment with my students at the start of each semester. We head into Washington Square in New York City, spread out, and try to blend in with the tourists, folk guitar players, and other students hanging out. I ask one of them to stand in front of the Washington Square arch in a dynamic pose with arms extended, and we all wait to see what happens as people walk by. This being New York, nothing happens – nobody even notices this oddly posed person standing stock still with outstretched arms. Then we repeat the experiment, except this time our "performer" wears a mask. Now everything changes: within a few seconds, people stop and watch and a crowd forms – now there is an audience waiting for something to happen. Our masked student does not move, and yet they wait and wait – surely there will soon be a performance to watch. This casual experiment shows that a mask worn by a person is a visual call for attention. It is a universal sign of theatricality, and it separates the wearer into the realm of drama, especially among the numerous distractions of an open-air venue. The simple act of donning a mask creates an instant theatrical frame. Here I am referring to Elam's concept of the theatrical frame, which is in turn influenced by Goffman's theories of frame analysis.[17] Whereas Goffman described the performative aspects of everyday public life and how people tend to adopt behavioral "frames" to mark off distinctive interactions, Elam applied this to the act of theatre "as a set of transactional conventions governing the participants' expectations and their understandings of the kind of reality involved in the performance."[18]

The Athenians' own myth of the origins of their theatre is about the very first time a mask created a performative frame, and this is intertwined with the creation of the dramatic character. In this tradition, an Athenian named Thespis (hence "Thespian") invented acting by masking his face with white lead powder and wine lees and in some sources inventing a linen mask.[19] He was said to have stepped out of a chorus in Attica and separated himself as an "answerer" (*hypokrites*) around the mid-sixth century BCE.[20] His act of masking separated him from both the chorus he had been part of and from his own persona as he acted out a completely new role; Thespis was no longer singing *about* a character, he was actually *being* the character. This story is probably apocryphal, although the tradition was certainly well known in the fifth century, as Thespis is named as the archetypal old-time poet in Aristophanes' *Wasps* and by Plato in *Minos* (322). Later, in Plutarch's *Solon* (29. 4–5), Thespis is associated with *kairos*, or novelty, which brings us back around to Aristotle's comments on crafting novel narratives and the mechanisms of predictive processing. Plutarch's biography also reinforces the connectivity of drama with political and social action as the elder statesman, Solon, is depicted fretting about the effectiveness of Thespis's new craft and worried that it might lead to people acting out falsehoods in their everyday lives. This is only a small step away from Plato's concept of the *theatrocracy*.

Although the type of dramatic mask we are concerned with here was probably specifically developed for the theatre, masks were not new in Greece, and evidence for them in cult practice and rituals date well before the mid-sixth century. This includes the votive masks found at the Sanctuary of Artemis Ortheia, near Sparta, dating from the mid-seventh century, mask depictions on *komos* vases, and evidence of Carthaginian masks found in Sicily, which date to the seventh and even eighth centuries.[21] Even older masks have been found in Israel, Egypt, and throughout the Near East. The earliest surviving reference to a mask in Greek is found on an Attic inscription dated to 434/433, which refers to a mask used in a ritual not associated with the theatre or Dionysos, its patron deity.[22] What is new about the Thespis myth is the way in which the first actor uses the mask both to separate himself from the chorus he was once a part of and to establish a character.

The affective mask

It has been shown that newborns are particularly sensitive to faces and respond to properly ordered faces over faces that have had their features re-arranged or have parts removed.[23] As early as nine minutes after a baby is born, a child will show attentional preferences for faces over other objects; at around 12 days old a baby is capable of imitating facial expressions; and by three months they can distinguish among different faces.[24] There is now a growing body of research that indicates that innate human facial perceptive skills contribute to both object and language processing, something we will explore further in connection with ancient Athens and literacy.[25] Additionally, the brain possesses a specialized neuronal network for face processing called the fusiform face area, located in the ventral stream of the temporal lobe on the lateral side of the fusiform gyrus, a brain region that is also involved in language recognition and production.[26] We might say that the face projects a grammar of non-verbal emotions and is what John Skoyles calls a "motor exposure board" in that it contains hundreds of muscles capable of generating a number of easily identifiable macro-expressions and a much larger number of seemingly imperceptible "micro"-expressions. When engaged in communication with another, the face is in a state of almost constant movement.[27]

Neurologist Oliver Sacks wrote: "It is with our faces that we face the world, from the moment of birth to the moment of death."[28] Yet, Sacks suffered from a condition called prosopagnosia or "face-blindness." This is an impairment of the ability to recognize familiar faces and evaluate emotional expressions, which denies a crucial visual mechanism of social interaction. Sacks said of his own prosopagnosia that "what is variously called my 'shyness,' my 'reclusiveness,' my 'social ineptitude,' my 'eccentricity,' even my 'Asperger's Syndrome,' is a consequence and a misinterpretation of my difficulty recognizing faces."[29] Difficulty decoding the emotional signals generated by the face is one of the major indicators for people on the autism spectrum, and it can be incredibly socially debilitating. As much as humans depend on facial processing to try to comprehend the thoughts and emotional states of others, a significant marker of normal cognitive development is our ability to display emotions that contradict our true emotional state. To know just the right time to feign a frown or a smile is a crucial part of normal human social interaction.[30] This is a basic aspect of mind/body dissociation connected with Theory of Mind and empathy.

With all of this in mind, how did the static mask with its ambiguous expression effectively communicate the wide range of affective states required of Greek drama in the expansive open-air space of the ancient theatre? Consider the mask of Oedipus in Sophocles' *Oedipus Tyrannus:* the text indicates that the character displays several emotional states in the course of the play, including confidence, anger, fear, confusion, and revulsion. How? In the play, the blind prophet Tiresias says he cannot be frightened by Oedipus's "angry glare" (448), yet it remains for the sighted audience members to visualize this look on the fixed face of the mask. Likewise, when the chorus of Aeschylus's *Agamemnon* sees Clytemnestra after she has killed Agamemnon and Cassandra, they proclaim her to be so crazed that they "can even see blood streaked in your eyes" (1428). However, I can think of no example from tragedy where one character says that they have read a changing emotion just from the facial features of another unless it is a reference to the eyes or tears.[31] There is of course one famous example of an expressive mask that is constantly cited, that of the so-called smiling mask of Dionysos in Euripides' *Bacchae*. Some have asserted that the Dionysos mask was presented with an unchanging smiling face based on two textual references, but the first is placed in the past (439), and the only other is a future hope expressed by the chorus (1021). There is no real evidence in the play that this mask was made with a fixed smiling expression.[32] This notion of the unchanging mask has dominated the thinking of classicists writing on the ancient theatre, encapsulated by Pickard-Cambridge, who wrote, "it was only facial expression that was unalterable, owing to the use of masks."[33] To compensate for this, he came up with some very odd "directorial" ideas about how Greek actors dealt with overt references to changing emotions in the text. His rather bizarre solutions include embraces that would hide the mask (and presumably the voice, making it inaudible) and having the actors constantly facing away from the spectators whenever something dramatic happened on stage!

Yet, most scholars have continued to assert that the Greek mask had a fixed expression or was neutral.[34] One can appreciate this premise if we observe the other figurative arts prevalent in the fifth century that survived – mainly sculpture and vase painting. When we observe the facial expressions depicted in the classical period, it is actually striking how rare it is to find human subjects with clearly defined expressions, except perhaps marginal figures such as centaurs, old women, and foreign subjects. The metopes of the Parthenon are emblematic of this aesthetic trait: the Greek male Lapiths show hardly any expression at all despite being punched, bitten, and trampled while the faces of the "uncivilized" centaurs are contorted in distinct expressions of pain, anger, or aggression. David Konstan has suggested that expression is minimal in these works because viewers draw their own emotional inference from the entire scene depicted and the gestures of the figures, which in many cases are certainly extreme.[35] This may be analogous to the mask in performances in that, as we shall see in the next chapter, the deprivation of the expressive human face actually enhances our perception of gestures, for more powerful and personal affective engagement. Yet there may be another aesthetic connection between the faces of Greek classical sculpture and the dramatic mask – the perceptual attributes of ambiguity.

The evidence shows that the "resting" expression of the mask could best be described as ambiguous and not neutral, and this helped it seem to transform in performance. Facial ambiguity can actually be an advantage when it comes to the rapid identification of characters, as some studies have shown. For example, Susan Brennan experimented with a schematic system where the face of a famous person was rendered as an exaggerated line-drawing caricature and set alongside an accurate line drawing of the real famous face. It was found that subjects strongly preferred the caricature face, which tended to be identified far more quickly than the realistic rendering of the same face.[36] As the art historian Ernst Gombrich noted, writing on what he termed the borderland between caricature and portraiture, "we generally take in the mask before we notice the face." Gombrich meant that a "mask" in this sense could be a caricature, or even a photograph, where the sitter's emotional state of mind and facial expression might actually communicate something entirely different in another contextual frame.[37] He showed that for portrait artists and photographers, the ambiguity of an expression is important, not neutrality, and it was the same with the classical mask. This ties in with Mori's theory of the "uncanny valley", a x/y axis chart that showed how people tend to prefer schematic versions of humans over ones that are too lifelike. The chart dips abruptly as the versions become too life-like and spikes when people saw a real human. Crossing this uncanny valley is still the goal of CGI (computer-generated imagery) filmmakers and digital game artists, and a human replica has not yet done it. However, Mori placed a Bunraku puppet across the valley and stated "although the puppets' body is not humanlike, we can feel that they are humanlike owing to their movement. And from this evidence I think their familiarity is very high . . . movement is a sign of life."[38] We will explore the cognitive impact of masked movement in the next chapter.

In many ways the Greek mask operated in a kind of cognitive gap by forcing the bottom-up/top-down mechanisms of the generative prediction model into attention and affective evaluation. Richard Gregory explained, "We not only believe what we see, to some extent we see what we believe . . . and the implications of our beliefs are frightening."[39] As mentioned in Chapter One, Gregory's famous demonstration of this was the hollow mask illusion. Here, a simple mask is lit normally and rotated slowly; as it turns to reveal the inside, the hollowed-out features suddenly seem to form into a three-dimensional face. The more realistic the face, the better the experiment works (Gregory used a plastic "Charlie Chaplin" mask from a joke shop). Even though the viewer knows that this is an illusion, it cannot be "turned off" as the mind's perceptual process still reconstructs the image of a convex face from a concave image. Gregory surmised that this was a biological function that had evolved as a protective measure to increase survivability.[40] Clark explains this strange phenomenon by relating it directly to the mechanism of prediction-based learning: "Our statistically salient experience with endless hordes of convex faces in daily life installs a deep neural 'expectation' of convexness: an expectation that here trumps the many other visual cues that ought to be telling us that what we are seeing is a concave mask." This is enormously advantageous "when the sensory data is noisy, ambiguous, or incomplete."[41]

Another reason for the lack of emotional faces in Greek art at this time may have been that the public display of emotions was considered by many to be unmanly and unseemly. One of Plato's criticisms of the theatre was that it encouraged public displays of emotions, and he notices that men will stem their emotions in public at a funeral but will not hesitate to burst into tears in the theatre (*Republic* 10.605c-e). Actually, as the theatre proliferated throughout the Greek world in the fourth century, it may have had a huge impact on the prevailing aesthetics promoting the expression of emotion in both art and public political speech. Such a view has been advanced by Angelos Chaniotis, who has identified a theatrical trope in the way in which ancient biographers described the use of emotional displays in political discourse.[42] There is also a corresponding change in the form and even the function of the dramatic mask, which I will briefly discuss at the end of this chapter.

Projecting emotions

Performing in a mask can be a difficult undertaking for most Western actors, who are often trained to "internalize" their own emotions and recall them in order to craft a "truthful" performance. Most mask work requires an opposite approach, where the external signs created by the actor create a kind of emotive choreography that is projected outwardly. In the 1970s John Emigh studied the masked performers of traditional Topeng drama in Bali. Topeng shares many of the same performative qualities as Greek classical drama, and when Emigh worked with Balinese mask practitioners they made it very clear that his facial expressions had to correspond to his body posture and the emotion he was trying to get the masked character to embody. He was told by his teacher that "the mask is not a disguise, if the face of the actor behind the mask did not register the character of the figure dancing, the body would move wrong and the mask would be denied its life."[43] Lada-Richards has described this kind of process as the creation of "a channel of communication between stage and auditorium sustained by the transfusion of emotion, the identity of shared feelings."[44]

David Wiles has proposed that the tragic mask transformed the performer and could hold power over the spectator who, in turn, projected personal conceptions upon the features of an almost blank mask.[45] This idea of personal projection is correct and, as will be described below, involved the same predictive perceptual process that trumps our vision in the hollow mask experiment. Wiles cites the work of Paul Ekman and his theory that six or seven facially displayed "basic emotions" are invariant across cultures and both commonly displayed and universally understood.[46] This has become a prevalent theory in neuroscience and face recognition studies, but is it accurate and can it be productively applied to the Greek dramatic mask?

The idea that there are facially displayed "basic emotions" can be traced back to Charles Darwin's seminal study, *The Expression of Emotions in Man and Animals* (1872). Since then, the questions of whether there are identifiable "basic" human emotions, if they are universal across human cultures, and what causes an emotion have become some of the most hotly debated premises in the study of human biology and psychology.[47] The usual premise of the various theories of basic emotions is that they developed as evolutionary biological social communicators and were then refined by cultural engagement. For example, the basic facial display of fear has a practical use: the eyes are open wide to receive maximum peripheral visual information, nostrils flare preparing to maximize oxygen intake, skin becomes flushed as blood vessels become oxygenated preparing the body for fight or flight, and the mouth is frequently open ready to express a cry of distress. According to most basic emotion theories, these physiological facial expressions can be quickly and easily identified by other people, even those from different ethnic, geographic, and cultural groups.

Ekman has proposed six or seven basic facially displayed emotions: happiness, sadness, anger, fear, surprise, disgust, and contempt (which is sometimes conflated with disgust). Ekman proposes that there are both "discreet" emotions, which are distinctive but do not necessarily require an evolutionary explanation, and "basic" emotions, which are universal across human cultures. It could therefore be ontology and not phylogeny that is responsible for commonalities in some emotional experience.[48] Ekman also proposes that his list of seven emotions may not be definitive, and further studies suggest that other emotions such as relief and wonder may also be universal. He has also pointed out that each emotion is not a single affective or psychological state but rather a family of related states, which he calls a "theme." According to Ekman, each theme is phylogenetically influenced, and the variations within each theme result from social experience. This permits variability, cultural differences in expression, and fluidity of display.

Jaak Panksepp takes a neuroscientific approach to the question of basic emotions and suggests a hierarchical system of emotional responses based on the architecture of the human brain. He names seeking, fear, rage, lust, care, panic/grief, and play as emotions that can be evoked by artificial activation of subcortical networks of the brain and therefore might be called "basic" and even universal.[49] Recently, new research has sought to revise Ekman's initial six or seven basic emotions down to four, suggesting that fear and surprise are congruent, as are anger and disgust. This research, led by Rachel Jack, suggests that four biologically rooted basic signals gradually evolved into more complex socially specific indicators.[50]

In my own work with masks, I ask spectators to watch an unmasked participant to silently display, with only facial expression and gestures, one of the basic emotions to the group. I then ask the group to tell me which emotion they see.

The answers are always diverse and the group never agrees. When the exercise is repeated again with a performer in a mask, the audience usually always agrees that they see "fear," "anger," "sadness," "joy," and "surprise." "Disgust," however, is far less clear. To enact these emotions, the actor is embodying the affective state in the body, with movement, gestures, and the angle of the head. Yet other "non-basic" emotions such as "jealousy," "love," or "envy" are almost impossible to convey non-verbally. Ekman, for example, considers jealousy to be an "emotional scene" with a plot of different emotions; he also claims that there is no evidence that jealousy is a unique distinctive universal signal.[51] What is clear from these exercises, which have been carried out with audiences from all over the world, is that without auditory context the embodied mask is fully capable of effectively projecting several commonly recognizable affective states.

The issue of whether certain emotions are fully recognizable across cultures is more difficult than perhaps Ekman, Panksepp, and other scientists allow, although I am persuaded that the ancient Athenians saw certain emotions in much the same way we do today. For the purposes of this study, what we can agree on is that the faces of all humans, wherever they are from, display emotions that are at least recognizable by other people within their own cultural group. Therefore, we do not need to see the same emotion in a face that an ancient Athenian did, to know that they also communicated emotions bodily to each other. However, like us, the Greeks were also concerned with categorizing emotions, albeit for specific purposes such as public speaking or artistic representation. For example, Plato lists pleasure, distress, confidence, and fear in *Timaeus* (69d); shame, love, and hate in *Laws* (1.674a-d); and anger, longing, jealousy, and envy in *Philebus* (47e). In *Rhetoric* (2.1–11), Aristotle lists anger, pity, and fear and describes them along with other emotions such as love, hate, shame, benevolence, indignation, emulation, and envy.

The theatre mask was a material object, not a human face, but a schematic and stylized representation of one: the product of artistic creation. The most detailed ancient reference to displayed emotions in art is found in Xenophon's *Memorabilia* (3.10.1–8). Here Xenophon describes Socrates visiting with Parrhasios the painter and Cleiton the sculptor and asking them how they went about achieving a lifelike quality in their respective art forms. Socrates asks Parrhasios if he is able to capture the "ethos of the soul" and lists a number of examples of good qualities such as lovability, friendliness, attractiveness, and desirability. Parrhasios responds that this would be impossible, as these qualities cannot be seen and, therefore, cannot be reproduced, either in form or in color. So then Socrates inquires if people usually express empathy and disgust by their looks, whether or not these feelings can be imitated in the eyes, and if so is it possible to make a copy of these expressions as well as the look of joy and sorrow. Parrhasios replies that this is entirely possible. Socrates adds to his list: magnificence, dignity, dejection, servility, self-restraint, prudence, insolence, and vulgarity, and says that they are all shown on men's faces and the aspect of the body whether they are still or in motion (*kai dia tou prósōpoou*). Parrhasios responds that he feels sure that these can all be imitated by art. Then Socrates visits Cleiton the sculptor and asks if the imitation of the emotions (*pathe*) that affect the body delight the spectator. When Cleiton responds in the affirmative, Socrates adds in that case the fierce look in the eye of a fighter should be copied and the look of pleasure in the face of a victor imitated. Cleiton agrees and Socrates concludes that the sculptor does represent the workings of the soul.

Socrates finds that a person's ethos can be visually replicated via mimesis and that their character and emotions "show through" (*diaphainein*) the face, eyes, and movements of the body.[52] This description could be just as aptly applied to the dramatic mask, a crafted object whose formation involved the plastic skills of the sculptor, the two-dimensional mastery of line and color of the painter, and the form-fitting expertise of the third artist Socrates visits, Pistias the armourer (3.109–115).[53] In fact, Socrates' description of the visible display of emotions to Parrhasios the painter (3.10.5) could also be equally applied to the mask. The four key terms – *prósōpon* (face), *schêma* (form), *stasis* (stillness), and *kinêsis* (movement) – are all essential elements that come together in the performer's mask and body to communicate the emotions of the character portrayed on stage. Xenophon's Socrates says that the face displays both a person's character and his or her emotions, but not in isolation – both face and body move together to convey this information.

This last point is vital in understanding how the mask operated in performance and communicated changing affective states. Just as Mori pointed out that the Bunraku puppet crossed the uncanny valley because of its lifelike movements, so the Greek mask was usually not presented as a disembodied object separated from the rest of the body. When it was, such as the visages that stun the satyrs in Aeschylus's *Theoroi* (*POxy* 2162), or as the inanimate head of Pentheus held by Agave in the *Bacchae* (1280–1290), the disembodied mask is an object of shock and dissociation. This highlights a problem with a good deal of neuroscience research carried out on the affective qualities of the human face. For example, a 2009 study by de Gelder found that 95% of social and affective neuroscience studies focused only on the face, whereas the remaining 5% most used scenes and auditory inputs, not the body.[54] The limitations of these studies are telling: most people on encountering a disembodied head in the world would be shocked to say the least. Just as we engage with the face as part of the entire body in the world, so ancient spectators engaged with the mask.

Externalist theories of distributed cognition and Clark's concepts of predictive processing all posit that the human mind is also constituted by components in the environment outside of brain, bone, and skin. Therefore, the body in the external environment is a fully functioning part of human cognition, and the face is part of that entire body system. Therefore, the mask, as a schematic surrogate face, cannot be understood in isolation detached from the body of the actor who was wearing it and the environment within which it performed. With this in mind, we will explore how the mask worked together with theatrical movement and gestures in the next chapter, here we will look at how the construction of the mask aided its affective transformations.

The mutable mask

The Greek mask was deliberately constructed to promote its affective mutability. Though no masks have survived from the fifth century, we can know something of their form and structure from the evidence of vase painting and sculpture, such as the Pronomos vase from the end of the fifth century, which depicts an entire theatrical cast in costume with masks (Figure 3.1). For the remainder of this

Figure 3.2 Terracotta figurine of a comic actor from Attica, late fifth/early fourth century BCE. Metropolitan Museum of Art, New York, Rogers Fund 1913. 13.225.20.

chapter, I am going to focus on the tragic mask, which had the same basic form as its cousin the satyr mask (Figure 3.3). The more bulbous comic mask (Figure 3.2) comes directly from the tradition of grotesque goblin-type masks used in the *komos* carnivals but also operated in the same way as its tragic counterpart. The Pronomos vase shows us a wealth of details about tragic and satyr play masks and costumes, and the artist has paid a great deal of attention to the masks, which are shown from a variety of angles. Let us look closely at the Herakles mask in particular (Figure 3.1). It is clearly a depiction of Herakles as indicated by the lion skin headdress and the facial features of a middle-aged man with copious hair and beard. The mask has large eyes, relaxed eyebrows, a furrowed brow, and an open mouth with a slight downturn. Now, compare this mask to the face of the actor holding it, and you will notice distinct similarities in facial features. Pickard-Cambridge also remarked on this uncanny aspect of the mask and suggested that the artist is depicting a "melding" between the mask and the actor.[55] In my experience this is a particularly "strange" quality of the mask – that when it is worn several times by a performer, it seems to take on that person's facial features. But when the same mask is given to another actor, it transforms and looks very different, even in photographs (Figures 3.4 and 3.5 show two photos of the same mask made in 2011 by David Knezz). This has a great deal to do with how we process

Figure 3.3 Red-figure chous or oinochoe fragment, c. 430 BCE. Courtesy of the Agora Excavations, American School of Classical Studies at Athens. (P. 32870).

Figures 3.4 and 3.5 Masks by David Knezz. Above, *Six Characters in Search of an Author*, the Father and Mother, Aquila Theatre, 2011. Below, the same mask as Amphitryon in *Herakles*, Aquila Theatre, 2012. Both performances directed by Desiree Sanchez. Credits: Renato Rotolo/Aquila Theatre and Miguel Drake-McLaughlin/Aquila Theatre.

the distinctive kinesthetic information of each performer's body, something I will return to in the next chapter on movement.

Look again at the Herakles actor's face and that of the mask he carries: the faces certainly resemble each other, but they are not the same; there are some important differences, and these differences provide some clues as to how the ancient mask was constructed and how it functioned. The most obvious difference is the size of the eyes. The mask has large, prominent sclerae (whites) and a purposeful direction of sight, which supports the idea of the frontal mask engaging the audience with its gaze direction. Look at how the painter has been careful to de-emphasize the expressive quality of the mask's eyebrows, one of the most animated parts of the human face, and paints the areas of expression – the corners of the mouth and eyes – as smooth surfaces with no attempt to denote expression of any particular kind. The mouth of the mask is slightly open, which is vital for audibility, but no distinctive emotion is shown. This is in contrast to the corner of the actor's closed mouth, which is turned up into a slight smile. The mask also has a high forehead with distinctive furrows, useful for catching light from the top and back and creating shadows, the play of which on the surface helped make the mask seem more animated. The actual expression on the face of Herakles is far from neutral, but it is ambiguous – we are unable to pin a distinct emotion on it, and this ambiguity helps gives the mask its transformative power. Also, notice the difference between the bottom lip of the mask and that of the actor: the mask's is large and protruding whereas the actor's is not. This same feature is found on Japanese Noh masks and is integral for facilitating their ability to seem to make affective changes.

After closely surveying the surviving evidence for fifth-century masks and noting these important differences between renderings of masks and faces in Greek art, I worked closely with a mask-maker, David Knezz, to develop working masks based on the form of those shown on the Pronomos vase. These masks (Figures 3.4 and 3.5) were then used in performance and exercises, in both modern artificially lit theatres and open-air ancient theatres in Greece. In undertaking that work, I found the following:

1   When combined with body movements, this type of mask is capable of showing at least five distinct non-verbal Ekman "basic emotions" by means of head and body movements: fear, joy, sadness, anger, and surprise. Disgust is also reported but less frequently, and it is often confused with other basic emotions; contempt is hardly ever recognized. More complex "emotional scenes" such as jealousy and love are not reported.
2   The expression on the mask must be ambiguous.
3   The corners of the eyes and mouth must be smooth.
4   The bottom lip needs to protrude.
5   It can greatly hinder the ambiguous quality of the mask if the mask-maker makes a determination of the emotional state of a character before making the mask.

The same structural distinctions found in Greek masks can also be seen in traditional Japanese Noh masks, and several studies have been undertaken to try to learn how its fixed face seems to be able to change its expression when placed at different angles in performance.[56] In one, a Noh mask was tilted in different directions and subjects were asked to report the expression they read in the face when it was placed at these angles (Figure 3.6). When the mask was tilted backward, most of the subjects saw happiness, and when tilted forward, the face appeared sad. One interesting facet of this study was that there was a marked divergence in interpretation between different cultural groups, with a Japanese control group reading different responses compared to the British group. Yet, both groups saw the mask change its emotions at the same time when manipulated, just different emotions. This might be explained by cultural factors, familiarity with Noh as an art form, and the difference between holistic and analytical facial processing between Japanese and British participants, which I will discuss later. Also, this experiment only showed participants photographs of the mask; it was disembodied, and therefore the viewer was unable to process any kinesthetic, gestural, or biophysical information from the actor's body, which are very important factors in mask cognitions that will be explored in the next chapter. Once again, I am not claiming here that we see the same mutable expressions of the Greek mask that the ancients did, but that it was capable of changing its expressions in spectators' minds, something that certainly does still happen today, wherever the viewers are from. Even a photograph of a disembodied mask Can stimulate the facial predictive processing of emotional projection (Figure 3.6). Interestingly, when a bare human face was placed at the same angles as the mask and shown to the participants, they all reported that its features did not change. As can be clearly seen from Figures 3.6 and 3.7, the change of the mask, even when disembodied, is quite dramatic.

The researchers of the Noh mask experiment also noticed how certain features of the mask were fashioned in order to assist with the mask being able to change

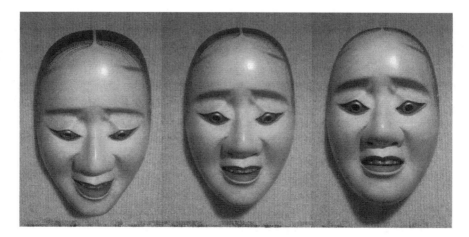

Figure 3.6 Three views of the same Noh mask under natural lighting. Only the tilt of the mask has been changed. Note the softening of the facial features and the curve of the mask. Credit: Wmpearl (public domain).

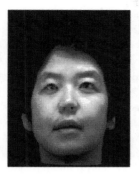  

Figure 3.7 Three views of a human face that demonstrate that it does not suggest different emotional states when viewed from a variety angles as the Noh mask does. From Lyons, M.J., Campbell, R., Plante, A., Coleman, M., Kamachi, M. and Akamatsu, S., 2000. "The Noh mask effect: vertical viewpoint dependence of facial expression perception" in *Proceedings of the Royal Society of London B: Biological Sciences*, 267.1459: 2239–2245.

its expressions depending on aspect.[57] For example, they noted how the mask's bottom lip was rendered to protrude and was much more exaggerated than the lips of a human face, just like the mask of Herakles on the Pronomos vase. Also, like the Greek mask, the Noh mask has smooth features at the corners of the eyes and mouth, where we tend to look for expression, and a large forehead with prominent eyebrows that resist emotional categorization. The Noh mask study also noted that minor movements of the masked actor's head could deceive spectators into thinking that the face is animated and the facial features are actually moving. From this comparative study, we can assume that the Greek mask, which shares so many of the same distinctive features with the Noh mask, also operated in a similar manner and was deliberately manipulated with head movements and body postures to provoke the facial and emotional processing systems of the spectator to perceive that the mask was changing its expressions.

This uncanny ability to project marked effective states onto the mask shows that even basic cognitive facial recognition systems quickly assess the threat level of that face by making rapid predictive guesses about the emotional nature of the face under view. Our top-down generative models have learned the kind of expressions, gestures, and movements that usually accompany the particular basic emotions we quickly tag. All it takes is the merest of movements of the mask for our predictive systems to interpret the bottom-up visual signals. As Figure 3.7 shows, the human face, when positioned at the same angles as the Noh mask, does not seem to change its expression, even in photographs. The schematic, visually reductive qualities of the mask act as a material anchor for the projection of our own best guess emotional predictions. Like the hollow mask experiment, the Noh and the Greek mask exist in the cognitive gap that is the moment-to-moment evaluative error-correction point between top-down prior models and new bottom-up sensory information. Thus, the mask is a prediction generator, and its spectator, prompted by the sensorial emotional signs produced by the actor (movement, gesture, speech, and song), which are often culturally specific, projects a deeply personal prediction of an emotion back onto the mask. In doing so it seems as if the mask has actually changed its expression. This is reflective of Merleau-Ponty's famous statement on facial perception: "I live in the facial expression of the other, as I feel him living in mine."[58]

David Kirsh maintains that this kind of enactive projection provides the foundation for sense-making.[59] Yet projection is not perception – it goes beyond perception to allow us to see what may not be physically present, but what *could* be – prediction again. Also, projection is not imagery, which requires no physical property but relies on working memory; rather, projection builds mental scaffolding on an existing material anchor. Kirsh calls this the "project-create-project cycle" – we project possibilities onto a material anchor to attempt to understand it, and these projections augment what we actually see. That mentally projected augmentation is then externalized, which "simultaneously changes the stimulus and makes it easier to project even deeper" (Kirsh 2009: 2312).[60] Such an externalization may be a doodle, a map, a tool, or in this case, a mask. Kirsh posits that these externalizations free up working memory and increase mental power. In the same way, the mask increases cognitive absorption and affective engagement. The Greek dramatic mask was a superb material anchor for affective projection.

Peripheral and foveal vision

The affective qualities of the mask were remarkable, but how could this type of mask, not much larger than the human face, communicate these emotional states in a large open-air space? An open-air performance is a very different experience from watching a show presented within an interior space. Distractions constantly compete with what is being presented on stage, whether the other spectators or the views available beyond the stage area. Open-air spaces tend to lack the kind of focus offered by modern proscenium or thrust stages, where the darkening of the auditorium and the framing of the performance space with the proscenium arch and modern theatrical lighting ameliorate the spectator's focus.

When we watch most modern plays, the actors are clearly within our central or "foveal" vision. This is named after the part of the retina that lies at the center and back of its curve. Human vision is bi-modal and oscillates between foveal and peripheral vision. Foveal vision is used for focusing on detail and scrutinizing objects, and peripheral vision orders the entire spatial view, allows us to look upon large items, and helps direct our narrower foveal vision. Neurobiologist Margaret Livingstone suggests looking at the world through a small tube, or our hands made into a telescope, to get an idea of how limiting foveal vision can be without the wider visual context supplied by peripheral vision.[61] Most modern theatre directors and designers work hard to keep our visual focus on the action they present on stage. Peripheral sights are normally regarded as distractions. This was certainly not the case in the fifth-century theatre. As we have seen, the visual environment of the southeast slope of the Acropolis held panoramic views of the sky, city, countryside, and sea, and the location was within the religious, civic, and cultural heart of Attica, with its important temples, shrines, statues, and civic buildings. Ancient dramatists were highly skilled in manipulating the dynamic interplay between peripheral and foveal vision in this space, and Greek drama offered a multi-layered visual experience that exploited the bi-modal capabilities of the human eye.[62]

Livingstone has provided an excellent demonstration of the interplay between foveal and peripheral vision in relation to the *Mona Lisa*'s famous smile. Leonardo deliberately blurred the expressive edges of his subject's mouth (*sfumato*), so our gaze is directed to fall between her face – in our foveal vision – and the landscape in the background – in our peripheral vision. Livingstone suggests gazing hard at the mouth and then looking at the background and then back at the mouth. As the mouth moves from our foveal vision to our peripheral vision, the expression seems to change. Thus, the viewer becomes intrigued and more engaged with this perplexing feature of the painting. Leonardo's technique pushes us beyond our normal visual expectations, exploiting the eye's duality of vision, and in so doing making us active spectators of his work of art.

Leonardo used his *sfumato* technique on the corners of the *Mona Lisa*'s mouth and the area around her eyes, knowing that these were the most expressive parts of the human face (at least for Western viewers). Likewise, the Greek mask was constructed with rounded features in these same areas. It was also worn within a performance space that was back- and top-lit by the sun. This would have cast gentle shadows on the features of the mask, which was constructed with dimensionality in mind.[63] Although the features were not exaggerated, the forehead, eye-sockets, eyebrows, cheeks, and lips were pronounced, and these were intended to assist the mask in seeming to change emotions.[64] In Japanese Noh theatre, the mask's ability to display a multiplicity of emotions is termed *mugen hyojo* ("infinite facial expressions"), whereas head and body posture and contextual factors such as vocalizations, music, and narratives all play an important part in this process, shadows are also an important factor in how the mask seems to change its expression.[65]

There are important differences between the staging of Noh drama and ancient Athenian classical theatre. Noh drama came to be performed in interior theatre spaces, although in the formative years of Japanese masked drama prior to the 15th century, this was not always the case. The Noh stage also used artificial lighting from both above and below, whereas on the Greek open-air stage the light source was the daylight. Sunlight in the open air does not function in the same way as the beam of an artificial light source. Natural light is non-specific and far more diffused, and it does not have the same kind of directionality as theatre lighting. Nevertheless, sunlight does cast shadows and changes in intensity and quality during the day as the sun moves from east to west.[66] In my own work with masks in ancient theatre spaces, I have found that the subtle interplay of shadows on the surface of the mask certainly seems to enliven its features. Shadows even suggest to the spectator that the facial features of the mask could be moving. In the aesthetic language of the Japanese Noh theatre, the concept of *yugen* (grace and subtlety) was regarded as the highest quality. This equates with the ambiguous quality of the facial features observed in representations of ancient Greek masks.

Furthermore, the neurons at the center of the visual process respond primarily to higher-resolution (fine) images, while those responsible for processing "the bigger picture" respond to images at a much lower (blurred) resolution (what is seen in one's peripheral vision appears blurred, until foveal vision is engaged to focus on the area). When low and high spatial frequency neural processing are combined – as with *Mona Lisa*'s undefined smile set against a distant landscape, or a tragic mask within an open-air setting – it can have the effect of confusing perceptual systems and causing facial features to seem to change.

In terms of emotional prediction, a study by Vuilleumier et al. has found distinct cortical pathways for sending visual information to the amygdala, the brain's emotional processing center. Participants in the study reacted more quickly and strongly to photos of fearful faces depicted in low spatial frequency or peripheral vision than in high spatial frequency or foveal vision. They propose that subcortical neural pathways provide the amygdala with coarse but rapid basic emotion information (they used fear), based on low spatial frequency features, independent of slower conscious analysis based on high spatial frequency cortical pathways.[67]

The mask operated in a bi-modal visual environment, its visage is already schematic and ambiguous, and this is enhanced by the subtle play of shadows across the masked faces of actors with the sun above and behind them. Although the Greeks' foveal vision allowed them to focus fairly well on the actors in the 6,000-seat fifth-century theatre, the features of their masks would have been deliberately harder to discern even by those seated on the front row. Add to this the bi-modal visual nature of the performance space, with the peripheral sights of the chorus, other audience members, cult sacrifices, city, countryside, and of course, the sky, and we can begin to understand how the mask operated in such a visually dynamic environment and would have been more successful than the actor's own face. The visual ambiguity of the mask complemented this environment in that its uncanny abilities to provoke the attention of the generative cognitive system forced individual spectators to engage with the mask on a deeply personal level, making their own evaluative judgments on the affective states their predictive system placed on the face of the mask. This was how the mask was able to create a distinctive kind of mimetic intimacy within such an expansive open-air space and why Thespis masked his face to become the first tragic actor; the mask fully engaged one of the most important and basic cognitive systems humans possess, the Theory of Mind ability to quickly predict and evaluate the emotional states of others, the most basic element in our ability to feel empathy.

Kuleshovian contexts

Just as the mask cannot be separated from the performer's body, in drama it cannot be perceived in isolation from the narrative context that surrounded it, or what is called "contextual framing." The narrative power of contextual framing and its effect on the ways in which facial features are interpreted was famously demonstrated by the 1917 film experiment of Russian director Lev Kuleshov. The so-called Kuleshov effect was used to demonstrate the effectiveness of film editing in creating emotional contexts that were then projected onto a neutral face. The same shot of the face of actor Ivan Mozzhukhin in the heavy white make-up of the silent film era is shown three times. In each shot (lasting around 3 seconds), the same facial expression is seen: a man staring intently ahead and then swallowing. Interspersed between this repeated facial shot are three different short scenes. The first shows a bowl of soup and then cuts to the face of Mozzhukhin, who seems hungry. Then the film cuts away to a shot of a child in a coffin, and when it returns to the actor, the same face now seems incredibly sad. The last scene is of a sexually desirable woman, and when the film cuts back to the face it now appears lustful. The purpose of Kuleshov's experiment was to demonstrate the power of a visually depicted situation to dominate emotional response. The facial expression does not change physically, and it is of course the same shot repeated three times, but our different emotional responses to the three scenes prime the way in which we view the meaning of the facial features.

Research on the contextual framing of faces and emotion has found that context is most potent in affecting emotional processing when participants viewed ambiguous faces, an important aspect of the type of ambiguous mask used in the fifth century.[68] Researchers at University College London replicated the Kuleshov experiment using functional neuroimaging on 14 subjects who viewed 130 facial images that were zoomed in and out on and juxtaposed with film clips in order to create a dynamic movement effect.[69] Fourteen supplementary images of humans, animals, and objects were also used to provide valence to the experiment. The results showed that faces paired with emotional film clips elicited strong neural responses in various regions of the brain that differed depending on the type of emotion shown. In particular, differing responses in the amygdala (a key part of the brain's limbic system responsible for memory and emotional processing) suggested that it acts to tag affective value to faces. Furthermore, the findings also suggest that a network of brain regions is deployed in "the storage and coordination of contextual framing." The anterior temporal regions store contextual "frames" (akin to top-down predictions), which are then compared to the information gathered by the superior temporal sulcas (involved in processing gaze direction and motion). These stimuli are then tagged by the amygdala, which is in turn influenced by top-down signals from the prefrontal cortex, the area of the brain responsible for executive functions.

This study offers a neurobiological basis for contextual framing effects on social attributions, and in so doing provides a glimpse into how the human brain operates when watching a mask in a drama. The brain's anterior temporal regions store known contextual frames, and the narrative of the plot offers such a contextual framing; the head and body movements of the masked actor are processed by the superior temporal sulcas, which, among many other tasks, is responsible for gaze direction and emotional processing. Remembered contextual frames provide top-down prediction models against which the visual sensory inputs suggested by the mask and body are compared. At this point the spectator perceives an emotional

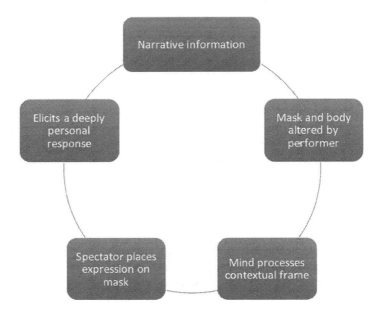

Figure 3.8 Neurobiological processing of the mask after Mobbs et al.

expression on the mask as suggested by its movements and other sensory contexts such as music, words and plot intentions, and surprisals. This information is then tagged by the amygdala, which helps regulate the neurochemical responses of the brain to emotions and is connected to working memory, this produces an embodied emotional response in the spectator (such as chills, increased heart rate, skin temperature, etc.). Then top-down signals from the prefrontal cortex, the area where the motor control and mirror systems are primarily responsible for executive function, elicit a deeply personal response to the suggested emotionality of the masked actor within the narrative context presented (Fig. 3.8).

The mask in its cultural and cognitive context

I have made some bold claims here for the power of the Greek mask and how it was fundamental to the development of ancient drama. If this is the case, why do most of us in the West not also respond to masks in a similar fashion when we see them today? There are many reasons we could cite for the mask falling out of theatrical usage in Western theatre: its condemnation by early Christians (along with the theatre in general), changing aesthetic tastes, a move toward theatrical realism that can be traced back to Elizabethan theatre, and the disentanglement of the act of theatre from cult and ritual practice. There was even a marked change in mask aesthetics in the ancient Greek world between the fifth and fourth centuries

BCE, which I attempt to explain later. Although we can view the evidence for Greek masks and even reconstruct them, we cannot view them within the same scopic regime or cognitive environment that the ancient Greeks did. We can know something about different attitudes toward the act of viewing found in ancient theories of extramissive vision and intraocular fire.[70] We can also discern how the notion of *miasma* or "pollution" meant that one could not look on the afflicted, and how in the visual culture of classical Greece looking was akin to touching in terms of social interaction.[71] Thus, the frontal mask with its large eyes gazing out at the audience may have been far more compelling to Greek audiences than it is to most of us today, especially as the masked characters were frequently enacting the breaking of one or more social and cultural taboos.[72] What is clear is that the ancient Greek audience looked at the mask differently than we do.

Cultural neuroscience can provide some new ways in which to approach this question. Up until fairly recently almost all studies assumed there was a universality of biological and evolutionary based brain function when it came to how humans perceived faces. This stemmed from research carried out in the 1960s by Alfred Yarbus, who developed a means of eye-tracking that followed exactly where the human eye scanned by recording the saccades (tiny flickers) of the pupil as it scanned faces. These highly influential studies found that participants looked at the eyes and the mouth, forming an eye-tracking triangle, and only then paid attention to the outline of the face and head. Subsequent eye-tracking studies on scenic vistas found that people focused on human forms when observing a scene before observing the background. When humans were not present, they looked first at the places where humans or other animals were most likely to be.[73] In Darwinian evolutionary terms, this kind of research seemed to make complete sense in that people needed to be constantly on the look-out for predators and to be able to quickly identify any potential threats. With this in mind, Yarbus's eye-tracking findings also indicated that humans looked first at the parts of the face that display emotions – the mouth and the eyes.

These kind of universalist theories began to be challenged in the early 2000s. A series of studies carried out by Takahiko Musada and Richard Nisbett found that eye tracking and visual processing actually differs quite significantly across cultural groups. One flaw with prior eye-tracking face processing studies was that they were measuring participants who were overwhelmingly from Western countries, whereas Musada and Nisbett compared participants from what they described as East Asian countries with those from Western countries.[74] According to their findings, the Westerners fixated on a focal object, analyzed its particular attributes, and sought to categorize the object, whereas the East Asians attended to a broader perceptual field, focusing on contextual information and causal attributions.[75] This was demonstrated in a number of visual recognition studies, including an experiment where participants from both cultures were shown a short video of a variety of animated fish swimming in a tank. When questioned on what they had seen, Westerners tended to first report on the salient objects, such as the size of the fish, how they moved, and their individual colors, whereas the East Asian participants reported the background, the color of the water, and the inert objects.[76] Nisbett proposed that prevailing cultural factors had affected visual cognitive mechanisms, with East Asians processing visuals more holistically, whereas Westerners were more analytical and relationship focused.

There is now brain imaging research that shows different neural networks operating between East Asian and Western participants undergoing the same cognitive activities.[77] Eye-tracking studies have also shown how people from different cultures look at artworks differently.[78] Studies like this, and other similar cultural neuroscience research, have posited that culture plays a central role in many aspects of cognition, including how we process faces and read emotions. This is because each distinctive culture has its own scopic regime, or way in which it perceives objects and surroundings. Humans all share the same biological functions for facial processing, and it happens in the same area of the brain, but the difference is the order in which things are processed and the cultural importance given to visual elements in the perception of the world around us. Neuroscientists have started to theorize on how and why different perceptual methods occur in different cultures. For example, Nisbett proposes that the particular nature of Chinese agriculture, involving mass cooperation, led to a culture where the collective was regarded as more important than the individual. According to him, this led to the development of holistic visual processing.[79]

Eye-tracking studies have now found that Westerners do tend to focus on the eyes and mouth, as Yarbus had originally concluded, but that East Asian participants tend to focus on the nose and center of the face. According to Caroline Blais, Rachel Jack, and their research collaborators, people from Western cultures move their eyes in a triangle between eyes and mouth looking for facially displayed affective information and building face recognition analytically, whereas in East Asian cultures, gaze aversion is a cultural norm, as is holistic visual processing. The center of the face at the nose is therefore the strongest single viewpoint to take in overall facial information (Figure 3.9).[80] Additionally, a recent study showed how when people looked at a central face placed in foveal view, which was flanked by other faces in the periphery, Australians spent more time focused on the central face, whereas Taiwanese participants tended to constantly track the faces on the edges.[81] There are also now some studies that suggest that initial instant facial tracking may be culturally invariant and is a secondary process of deeper interaction dependent on the prevailing culture.[82] Nevertheless, eye-tracking studies have led the way in showing how culture affects cognition.

We can apply this research to ancient evidence. Even taking the vagaries of vase painting into consideration, in the fifth-century representations of masks we do see some distinctive commonalities. They all have a relatively small nose, large eyes, and a normal-sized open mouth. The facial features are smooth around the eyes and mouth, and the bottom lip of the Greek mask tends to be more pronounced. If we compare the Greek masks to Japanese Noh masks dating from the 15th to 19th centuries, we immediately discern some marked differences. The Japanese masks all have very large noses that dominate the mask's face, relatively small eyes, and stylized mouths. Even on the grotesque "demon" masks, with their gaping mouths and deep-set eye sockets, the nose is by far

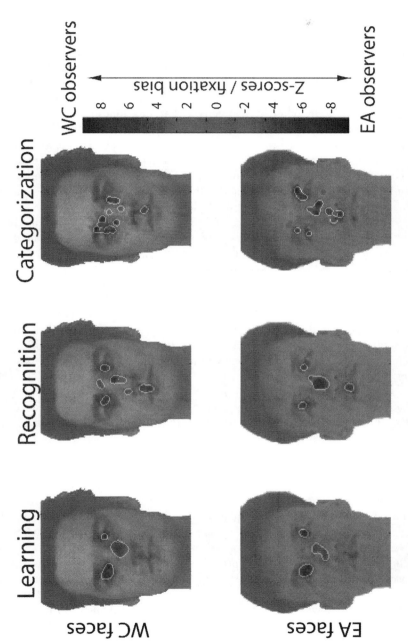

Figure 3.9 Fixation biases for Western Caucasian (WC – red; see online, www.routledge.com/9781138205529) and East Asian (EA – blue) observers. Blais C, Jack RE, Scheepers C, Fiset D, Caldara R (2008). Culture shapes how we look at faces. *PLoS ONE* 3(8): e3022. doi:10.1371/journal.pone.0003022.

the most prominent feature, and the eyes are actually fairly small. I suggest that what we are seeing here is not only the aesthetic preferences of the two different cultures but also a physical mimetic representation of their respective cognitive facial processing systems. From the evidence of the Greek mask iconography, it seems at first sight that the ancient Athenians processed faces much as Westerners do today, focusing on the eyes and mouth, whereas Japanese audiences for Noh drama processed faces in much the same way as Nisbett's modern East Asians, by focusing on the nose in the center of the face, hence the oversize, dominant noses on the masks. It is also notable that the Chinese character that denotes "self" (*ziran*) is made up of the word for "nose" (*zi*) and the word for "correct" or "yes" (*ran*). Furthermore, in Chinese cultural practice, the term *zi* is also used to denote a personal point of view, and many people point to their noses to indicate themselves.[83] Thus, in the facial features of mask representations, we may well be able to discern some of the cognitive qualities of ancient Greek facial processing.

Does this then mean that the Greeks viewed the face and therefore the mask in the same way we do? In some respects, perhaps: following Nisbett we could argue that we have here evidence for a more field-independent culture, and that would certainly fit with the cultural and political life of democratic Athens and the newer focus on individuality (the actor as "invented" by Thespis), whereas an older cultural tradition of collective expression still exists (the chorus). For example, Elsner has commented on a visual turn in Athenian culture around the end of the sixth/start of the fifth century, and Farenga has traced a similar shift from group to individual in politics.[84] What has been identified at this time is a shift in cognition because of rapidly developing cultural factors. As Elsner comments on the development of individual subjectivity,

> the viewer ceases to be the direct recipient of a poetic declamation and instead becomes an observer of a world imaginatively constructed to be like his or her own world, to which the viewing relationship is one of indirect identification and imaginative absorption.[85]

The relationship between the chorus, the audience, and the actors epitomizes this cognitive shift in mimetic form, as does the mask of the individual actors operating in a field of bi-modal vision where peripheral and foveal vision operate together offering shifting perspectives among chorus, actors, and environment. One of the other significant cultural changes that came as a result of these cognitive shifts was a marked rise in textual production and literacy, and recent studies in neuroscience can help us understand more about how the Athenians may have perceived the mask differently from us today.

With all this in mind we cannot assume that the ancient Athenians looked at the mask quite the same way that we do today. Here we need to consider the cultural differences between us and the ancient Greeks and how they might have affected cognition and visual perception. The example I will briefly explore is the impact of literacy and how literate people perceive faces differently from semi-literate and illiterate people.

Culture, literacy, and face processing

Brain imaging studies have indicated that the processing of faces is performed primarily by what has been named the fusiform face area, part of the brain's fusiform gyrus (ridge) located on the lower part of the temporal lobe between two other long ridges, the occipitotemporal gyrus and the parahippocampal gyrus. This area is associated with high-level image processing and object recognition. This hypothesis was dramatically confirmed by a 2012 Stanford study carried out on a live human subject who had electrodes placed directly on this part of his brain.[86] This was possible because the man in question was undergoing electrocorticography (electrode stimulation of exposed brain tissue) for acute epileptic seizures and had a section of his skull removed exposing the area where the fusiform gyrus is located. When electrodes where attached to two nerve clusters about half an inch apart (designated pFus and mFus) and stimulated with an electric current, temporarily lobotomizing them, the patient reported that his doctor's facial features "melted away." When the current was stopped, the patient was able to recognize faces instantly.

The fusiform face area shares a good deal of architecture with the fusiform gyrus, in particular what has been called the visual word form area (VWFA), which is associated with reading. This area is left lateralized in most people and shares networks with the areas of the brain responsible for language processing in the left superior temporal and inferior frontal regions. It has been shown that the neurons within this ventral visual pathway tend to respond to basic intersecting forms such as "T", "l", and "X" shapes. Recently, Stanislav Dehaene has proposed that all writing systems use variants of these intersecting contoured shapes, to which our primate brain was already "highly attuned." This helps answer Dehaene's basic research question: how did humans become so incredibly efficient in reading and writing, when writing systems are no more than 4,000 years old, which is not nearly enough time for the brain to have evolved complex reading functionality on a Darwinian evolutionary model?[87] Dehaene posits that the brain is capable of quickly "recycling" the existing neural networks that originally evolved to process similar, but different tasks. In this case, he suggests that when people learn to read, they develop a neuronal "short circuit" between the left ventral visual pathway and the left-hemispheric language areas. In this model, human cultural expression can never be independent of biology, and cultural inventions, such as writing, can only happen because they utilize preexisting neural architecture. Dehaene calls this "living proof that culture is constrained by brain biology."

Antiquity provides an example of the kind of illiterate cognitive processing Dehaene has described. A fragment of Euripides' *Theseus* dating to the end of the fifth century includes the character of a herdsman who is reporting (possibly to King Minos) an inscription that reads "Theseus" and describes the letters he has seen in terms of bisecting lines and curves that resemble naturally occurring objects:

> I don't know how to read or write,
> But I'll describe their shapes and offer you a clear account.
> There's a circle neatly measured out, as if turned on a lathe
> with a prominent mark in the middle. [θ]
> As for the second letter, there are two lines, first of all,
> And one more in the middle that connects them. (E)
> The third resembles a curling lock of hair; (S)
> as for the fourth, one line stands up straight,
> and three crooked ones are propped up against it. (E)
> The fifth letter's not easy to describe:
> there are two lines that are separate from one another,
> although they merge into a single base. (Y)
> And the last letter's like the third. (S)
>
> Euripides, *Theseus* (*TrGF* 382 = Athenaeus
> 10.454b-c), translated by R. Gagné

Euripides' Herdsman is clearly illiterate, and this has an important impact on the way he would have perceived a face. Dehaene has shown that the acquisition of literacy skills involves a reconfiguration of existing brain functionality that creates a lasting difference for facial processing. To test his hypothesis of neuronal recycling, Dehaene and his colleagues tested the brains of illiterate people, comparing them to literate and semi-literate people. What they found was that when participants were shown a string of letters, the visual word-form area became highly activated at the left lateral occipitotemporal sulcus, within the VWFA. The level of activation was in direct proportion to the participant's level of literacy – the more literate the person, the greater the activation.

The fusiform face area lies within the same region of the brain, and Dehaene found that in the illiterate subjects, faces and objects provoked intense activation in this region, but that the activation actually decreased the more literate the participant was. Yet this activation was only found in the left hemisphere; in the fusiform face area of the right hemisphere, activation to faces actually increased with literacy. Dehaene concludes:

> The acquisition of reading seemed to induce an important reorganization of the ventral visual pathway, which displaces the cortical responses to the face away from the left hemisphere and more toward the right. This displacement is presumably because the features that are most useful for letter recognition (configurations and intersections of lines) are incompatible with those that are useful for faces, so that one pushes the other away.[83]

In another recent related study, it has been shown that illiterate people process faces far more holistically than do literate people, who tend to be much more analytical and adept at recognizing details.[89] The study concluded that brain reorganization induced by literacy reduces the automatic holistic processing of faces. If we link this important research to the 2012 Stanford "face melting study," we might theorize as to how the brain utilizes the left and right fusiform face areas. In the Stanford study, when the participant had two nerve clusters in his right fusiform gyrus stimulated by electricity, he told the researcher who he was looking at, "your face metamorphosed" and "you almost look like somebody I've seen before, but somebody different."[90] Yet, the faces of pictures of famous people he observed did not change when the right fusiform gyrus was stimulated. We might conclude from this that the left fusiform face area evolved to quickly identify a face and process it holistically, to assess threat levels (the reason why we are prone to see faces in the clouds, or the face of "Jesus" in a tortilla, etc.), whereas the right supplies contextual information to aid categorization such as gender, age, distinguishing marks, etc. The more that reading skills are developed, the more activity to faces is seen in the right hemisphere, while the left diminishes. This is also supported by recent studies showing that many children with dyslexia process reading in the right hemisphere, whereas non-dyslexic people process reading in the left hemisphere.[91]

To put it simply, both illiterate and literate people react quickly to what they perceive as a face in the visual environment. A literate person who has reorganized neural networks to be able to read has an increased ability to call on contextual and analytical cognitive tools, but an illiterate person tends to retain a stronger, more intensely emotional response to the face for a longer period. Dehaene points out that in "recycling" brain function we can perceive the changes in process. For example, he cites the way in which young children learning to read will confuse "b's" and "d's" as they teach their ventral pathways to disable the symmetrical processing systems designed to deal with naturally occurring objects.[92]

With a neural explanation for differences in face processing across literate abilities in place, we can now apply the science to what we know about levels of literacy in fifth-century Athens. Rosalind Thomas has shown that there were different literacies operating in fifth-century Athens and that the number and ability of people to read accelerated as democracy broadened and more state business was enshrined in writing.[93] Even so, Thomas has suggested that "functional literacy" was at best extremely basic in the mid-fifth century. Anne Missiou broadly agrees with this view and has posited that it was still possible to function quite well within fifth-century Athenian society with either very limited or even no literacy at all.[94] There are, however, a number of references to reading and letters in Greek drama, but Carl A. Anderson and Kenneth Dix have argued persuasively that the several references to reading and papyrus scrolls found in Aristophanes may actually portray a suspicion of literacy, as if people pretended they could read but really could not. The availability of papyrus scrolls was in many ways a new technology that only a few members of Athenian society had mastered. It was the cusp of a communication revolution, not unlike the advent of personal computers in the 1970s and early 80s or the mysteries of the Internet and e-mail to most people in the early 1990s.[95]

Within Greek plays, we can detect a distinct cognitive shift toward the written word in Athenian culture. Isabelle Torrance has shown how the implements of writing do not appear in any of the extant plays of Aeschylus and Sophocles and that all references to writing are related to the recording of memory. In this way, writing is perceived as the recording of an important speech act that is enacted again by speaking the written words. This all changes with the plays of Euripides, where letters and inscriptions become agents within the plot structure of the plays, and "for the first time in antiquity writing is internalized and processed as *text*, rather than being read aloud."[96] Torrance concludes:

> the different ways in which the three great tragedians explore the concept of writing . . . suggests a marked development in the perception of writing over the course the fifth century, and suggests that only the elite had the kind of literacy skills necessary to being capable of reading the text of a play script.[97]

In the fifth century, the Athenian elite strove to be literate, composing speeches and controlling the codifications of statutes, laws, and trade accounts. The average Athenian citizen probably did not need more than a basic recognition of letters to actively participate in Athenian society, which Thomas suggests might explain the preponderance of lists at this time.[98] The next logical question then is who was the audience for Athenian drama – the elite or the same kind of Athenian citizen who attended the assembly and law courts? New archaeological research on the theatre of Dionysos suggests that the fifth-century *theatron* sat around 6,000 people.[99] Although not able to accommodate the 30,000 or so Athenian voting male population (the *demos*), this was still as large a public space as the Pnyx. This was the site of the Athenian assembly, which also accommodated around 6,000 and by most ancient accounts included a cross section of the Athenian *demos*. From the wise-cracks of the ribald characters of Old Comedy and the political themes lampooned in the plays, we can assume upper-, middle-, and lower-class Athenian citizens were all present at the theatre, along with a number of invited ambassadors and dignitaries from allied states. David Roselli has even suggested that non-Athenian citizens such as women, foreigners, and slaves could have conceivably also been watching the plays.[100] We can therefore assume that the majority of the spectators were functionally, but not fully, literate.[101]

Yet, even among the Athenian elite there may have been an ambivalent attitude toward literacy. Plato was famously suspicious of writing, believing that it would erase memory, that it was only the semblance of wisdom, and that by using writing to teach knowledge, people would end up filled with the conceit of wisdom but would actually know nothing (*Phaedrus* 275a-b). A famous anecdote found in Plutarch (*Nicias* 29.2) may shed some light on this question as it relates to drama. He writes of a few Athenians who were able to pay their way out of the Sicilian quarries and get home to Athens after being captured in the disastrous Sicilian expedition. This was because they could sing the choruses of Euripides. That Sicilians would pay to hear Athenians come to their homes and sing from the plays suggests that the play script as a transportable text had not yet emerged and oral transmission was still optimal. Also, when we find depictions of dramatists in comedy, they are portrayed composing rather than writing, and there is a mention in Plutarch of Euripides teaching his chorus his lyrics by call and response.[102] Thus, throughout the fifth century the spoken or sung word was still the most effective means of communicating drama, and literacy was probably not required to either perform in the plays or watch them.

To sum up, illiterate and semi-literate people process faces more holistically and contextually. They tend to rely on more instinctual and emotional processes to parse faces, over analytical cognitive tools. If the fifth-century audience members were predominantly semi-literate or illiterate, then their perceptual responses to the mask would have been significantly different from that of most modern Western theatergoers in that they were more intensely emotional and far less analytical. Perhaps this is why the mask is regarded so differently in the West today, where it has become the preserve of the carnival, the mime, the disguised protester, comic superhero, or a child's toy. For most of us in the West, masks simply do not provoke the same neural responses that they did for the Athenians. Then what about the continuing traditions of masked Noh and the use of mask-like heavy make-up in Kabuki theatre and Chinese opera? Nisbett's East Asians are from literate cultures, and yet he found these participants to also be more inclined toward holistic and contextual visual processing. There has also been a good deal of research on the cognitive differences between language systems showing that Westerners' linguistic reasoning is relatively analytic, which includes the tendency to focus on categories, whereas Chinese is relatively holistic, including a tendency to focus on relationships.[103] The fact that masks still play an important role in East Asian cultures is evidence that societies that are more field dependent tend to also be visually holistic when it comes to face processing.

Neural networks cannot tell the whole story, and this is where the generative model of predictive processing becomes useful. Theories of distributed cognition, which Clark incorporates into his model, place the mind and the body in the world, and the cognitive feedback loop between them is continual and fluid. Nisbett is probably right that the collective farming practices of East Asia, which developed to suit that particular geophysical environment, has profoundly shaped local cognitive processes. In the same way, the Greek landscape, which is peppered with small plains and divided by low mountain ranges, helped foster independent ancient city-states. Thus, as polis organization started to replace the older hierarchical aristocratic social structures, so we also perceive a corresponding cognitive shift where collectivity is blended with individuation. The mask functioned within this developing scopic regime, both as emotionally affecting, as supported by the responses of the more holistic chorus, and able to project the intentions and personal emotions of the individual character.

A cognitive change of masks

A final but important point: I have suggested that the Greek tragic mask (the comic version developing independently from the ribald traditions of the *komos*) developed as a reflection of changing cultural practices in Athens that in turn had an effect on cognitive processes. If this is the case, and the theory that theatre practice reflects the cognitive regime of the culture it is performing for is correct, then we should see the mask change also as literacy levels rose. Chaniotis and Rosenwein have both independently advanced theories that any history of emotions should include significant changes within the culture under scrutiny.[104] There was certainly a dramatic rise in levels of literacy in Greece from the end of the fifth century into the fourth. The Ionic alphabet was officially adopted by the Athenians in 403, and in 405 Aristophanes depicts his comic Dionysos reading a text of a Euripides play, *Andromeda*, while on board ship in *Frogs* (52–54).[105] There is also a tantalizing fragment of a comedy by Kallias called the *Alphabetic Tragedy*, which has the 24 members of the chorus resemble the 24 letters of the Greek alphabet and sing and dance a phonetic farce.[106] Was this lost play originally a comic commentary on a recent spike in literacy education? Books were certainly starting to proliferate, and in Xenophon's *Anabasis* we learn of a large cargo of books that were part of a shipwreck in Northern Greece (7.15.14).[107] By 330 BCE, the Athenian statesman Lycurgus ordered an official set of dramas to be written down and lodged at the Athenian state library to be preserved for posterity. We also see a rise in the founding of polis libraries, a growing fashion for the elite to collect books, and written texts on papyrus being used in schools throughout the Hellenistic Greek world.[108]

At the same time, we can discern a noticeable change in the Greek dramatic mask, especially for tragedy. The tragic mask of the Hellenistic period became highly stylized, larger than life, and elaborately decorated. The functionality of the mask changed, as had the Hellenistic theatre. The role of the chorus had diminished and all but disappeared, and leading actors became internationally famous. The theatre spaces had grown in size and scale, as did the mask with a high headdress and a large gaping mouth and eye-holes. These features allowed the audience to be fully aware of the actor behind the mask. It was also more formalized stylistically and less able to seem to change its emotional states. By the third century BCE, we find a number of depictions of masks rendered to show the eyes and mouth of the actor behind the mask.[109] The Hellenistic mask called attention to itself as a mask – it was never intended to become the face of the actor wearing it as its fifth-century predecessor did – instead it was an object of design that signaled an idea of classical drama – a decorous prop for a literate audience.

In this chapter I have attempted to re-conceptualize the Greek dramatic mask as a dynamic emotional motor exposure board capable of eliciting deeply personal affective responses from the spectators at the theatre. I have shown how the mask was not neutral and did not project just one fixed expression as others have argued, but was instead ambiguous, schematic, and a material anchor for affective predictive projection. I have also analyzed the form and construction of the mask and found effective parallels with the traditional Japanese Noh mask. Finally, I argued that the Greeks viewed the mask quite differently from us and would have had a more profoundly emotional experience watching it. A number of times in this chapter I alluded to how the mask did not work in isolation and that a large part of its success was its ability to focus kinesthetic empathetic attention to the body and emphasize movements, gestures, and even spoken and sung language. These crucial affective elements will be the subject of the next chapter on movement.

Notes

1   Such as, Easterling 1973, 1977, and 1990; Gould 1978; Goldhill 1990; Gill 1996; Grisolia and Rispoli 2003; Thumiger 2007; Seidensticker 2008; Budelmann and Easterling 2010.
2   Gould 1978: 44.
3   Marshall 1999: 189.
4   Wiles 2007: 169.
5   Seidensticker 2008: 341.
6   Aristophanes *Birds* 674: taking off a mask is compared to peeling an egg; *Knights* 230–232: the mask-makers are too scared to make a mask that resembles Cleon; fr. 31 (Henderson), Scholium on *Peace* 474: a reference to hideous goblin masks of Mormo; Fr. 130 (Hederson:– a reference to *mormolukeia* (goblin masks) being hung on display at the Sanctuary of Dionysos; Cratinus *Seriphioi* fr. 218: may be a reference to a tragic mask being handed to a comic Perseus. See Bakola 2010: 159–160; Aeschylus *Theoroi* (POxy 2162) may also contain a reference to the masks of the satyrs that are hung up on a temple wall. See Taplin 1977: 420–422.
7   Pickard-Cambridge et al. 1968: 187; Brockett et al. 2016: 27–20.
8   See Griffith 1998: 244–245; Halliwell 1993: 200, n. 11.
9   Hall 2012: 29.
10 Jones 1980: 43–44; Wiles 73 & 153–154; Vovolis and Zamboulakis 2007.
11 At the ancient stadium at Delphi at the Delphi Festival in 1990; Epidauros in 1992; Athens in 2007; Argos and Nafplion in 2011. In Argos, the masks used were specially made by David Knezz based on the author's research (see figures 4 and 5).
12 Jones 1980: 44. On Jones's impact, see Wiles 2007: 275–277.
13 Halliwell 1993: 211.
14 Calame 1986: 125–142, and Calame & Burk 2005: 113–114 and 119–123.
15 Meineck 2011: 121–124.
16 Apart from a fleeting reference to a comic mask in *Poetics* at 1449a37-b1.
17 Goffman 1974.
18 Elam 2002: 79.
19 For a detailed analysis of the sources for Thespis, see West 1989.
20 *Suda* θ 282; *Suda* θ 494; Athenaeus 14, 622c., Calame 1986: 126, n.6 and Csapo and Slater 1995: 89–102.
21 Nielsen 2002: pl. 43.
22 *IG* 13 343.7.
23 Chambon et al. 2008: 73–76.
24 Simpson et al. 2014.
25 Engelmann and Pogosyan 2013.
26 Wiggett and Downing 2008: 420–422; Rossion et al. 2012: 138–157.
27 Skoyles 2008.
28 Sacks 2010: 82.
29 Sacks 2010: 85.
30 Huang and Galinsky 2011: 351–359; Rivolta 2014.
31 See Griffith 1998: 244–245, who lists references in tragedy to eyes, tears, eyebrows, some to hair color, and a few to lips (where foaming at the mouth is a sign of madness as at Euripides' *Heracles* 934). Pickard-Cambridge lists examples of characters crying in tragedy; Pickard-Cambridge et al. 1968: 171–172.
32 Foley 1980: 107–133. This idea of the "smiling mask" of Dionysos was also put forward by Dodds 1944: 131. Marshall 1999: 196 also feels that "it is not self-evidently

true." For discussion, see Wiles 2007: 220–222, who also questions the notion of a "smiling mask."

33 Pickard-Cambridge et al. 1968: 171–174.
34 Jones 1980: 44. For Halliwell, the masks radiated the "ethos of heroic dignity in the midst of destructive settings"; Halliwell 1993: 211. The issue of aesthetics and the mask are also discussed by Hall, who is more sympathetic to the notion of a mask that operates within the consciousness of the spectator but still seems confined to the idea that the mask offered just one fixed expression; Hall 2009: 99–141. Wiles 2007: 205–223 calls the tragic mask "Idealized." McCart 2007: 247–257 has conducted experiments with masks and concludes that they are not neutral but are combined head and body movements (*schemata*) to become "a powerful dramatic tool."
35 Konstan 2006: 30.
36 Brennan 1985: 170–178; Benson and Perrett 1994: 5–93.
37 Gombrich 1973: 112–118.
38 Mori 1970: 34; see also Mori et al. 2012.
39 Gregory 1970: 11.
40 Gregory 1997: 1121–1127.
41 Clark 2015: 50–51.
42 Chaniotis 1997: 219–259.
43 Emigh 2003: 110.
44 Lada-Richards 2002: 414; see also Lada-Richards 2005.
45 Wiles 2007: 225. Csapo 1997: 253–295 holds a similar view: "Masks are the concrete embodiment of the power of Dionysus, because Dionysus works his particular magic through possession, especially through the eyes, creating a kind of enthusiasm in the etymological sense of *entheos*, the god being inside one."
46 Ekman 2007.
47 Plamper 2015: 144–163.
48 Ekman and Cordaro 2011: 364–379; Ekman 1999: 308–320: Ekman 1999: 45–60.
49 Panksepp and Watt 2011: 387–396.
50 Jack 2013: 1–39; Jack et al. 2014.
51 Ekman 2007: 366.
52 Halliwell 2002: 122–124.
53 See Hall 2006: 103.
54 de Gelder 2009: 3475–3484.
55 Pickard-Cambridge et al. 1968: 187.
56 Lyons et al. 2000: 2239–2245.
57 Another team in Japan also performed a similar test on facial recognition and emotions of Noh masks and found similar results. See Minoshita et al. 1999: 83–89.
58 Merleau-Ponty and Toadvine 2007: 174.
59 Kirsh 2009: 2310–2315.
60 Kirsh 2009: 2312.
61 Livingstone and Hubel 2002: 69–71.
62 See Sourvinou-Inwood 2003: 22–25, where she describes the perceptual frame of Greek drama as "zooming" between the mythic past and contemporary religious practice.
63 Hall makes the point that "the facial contours of the masks in tragedy seem to have been softly rounded, rather than using sharp angles and planes to represent three dimensions"; Hall 2006: 101.
64 This ability of shadows to affect the visual qualities of the mask has been pointed out by Pat Easterling. See Easterling 1997.
65 Kawai et al. 2013.
66 Kleffner and S. Ramachandran 1992: 18–36.
67 Vuilleumier et al. 2003.

68 Wieser et al. 2014: 74–82.
69 Mobbs et al. 2006: 95–106. See also Halberstadt et al. 2009: 1254–1261, where findings from electromyography (EMG) strongly suggested evidence of concept-driven changes in emotional perception, highlighting the role of embodiment (the role that the body plays in shaping the mind) in representing and processing emotional information. See Hietanen and Astikainen 2013: 114–124 and Van den Stock et al. 2014.
70 Several pre-Socratic thinkers, such as Alcmaeon of Croton, Empedocles, and Democritus, held theories of intraocular fire, and even Plato explained that the "pure fire within" is caused to "flow through the eyes" where it coalesces with the daylight before it reaches the object (*Timaeus*. 45b-46c). See Lindberg 1976: 1–17; Wade and Wade 2000: 11–19.
71 See Blundell et al. 2013: 15–16 and Rudolph 2015.
72 On the frontal gaze and Dionysos, see Csapo 1997: 256–258. On vase painting, see Korshak 1987.
73 Yarbus 1967. See also Tatler et al. 2010.
74 Masuda and Nisbett 2001.
75 For a critique of Nisbett's oversimplification of his participants' ethnicities, cultures, and personal experiences, see the excellent discussion by Geoffrey Lloyd, who calls them "global hypostatizations" in Lloyd 2007: 160–170.
76 Nisbett and Masuda 2003: 11163–11170.
77 Han and Ma 2014.
78 Liu et al. 2013.
79 In cognitive tests Nisbett suggested that East Asians are more "field dependent" than Westerners in the way they viewed a landscape or even a simple schematic called the "Rod and Frame Test," where the participant sits in a dark room and is asked to lay a luminous rod vertically within a slightly leaning luminous square frame. Field-dependent people tend to line the rod up with the edge of the square and not lay it vertically, whereas field-independent people (most Westerners according to Nisbett's study) lay the rod vertically, ignoring the square. Field-independent people tend to perceive their own bodies in relation to the rod, not in relation to the surrounding environment (the square); Nisbett 2004.
80 Blais 2008.
81 Lee et al. 2016; Ji and Yap 2016: 105–111.
82 For example, Or et al. 2015: 12–12.
83 Callahan 1989: 171–189.
84 Farenga 2014: 84–100; Elsner 2006.
85 Elsner 2006: 88.
86 Parvizi 2012: 19–40.
87 Dehaene 2013.
88 Dehaene 2013: 9.
89 Ventura et al. 2013: 105–109.
90 Parvizi 2012: 14915–14920.
91 Raschle et al. 2012: 2156–2161. See also Wolf 2012.
92 Dehaene 2013: 10–12, See also Blackburne et al. 2014.
93 Thomas 2001: 68–81.
94 Anne Missiou 2011.
95 Anderson and Dix 2014.
96 Torrance 2013: 142.
97 Torrance 2013.
98 Thomas 1995: 59–74.
99 See Chapter Two, page 59.
100 Roselli 2011.

101   Pappas has identified a tension in fifth-century Athenian political life between those
        who could read and those who could not, and has suggested that the presentation
        of writing in the theatre was accompanied by visual markers and staging devices
        that made the meaning accessible to both "educated and uneducated alike"; Pappas
        2001: 48.
102   There are references to the act of reading in tragedy, but this is not evidence for full
        literacy. Aesch. fr. 358 is advice to a warrior to fight at close quarters, "like a man
        reading in old age." See also Sophocles fr. 858.
103   Imai et al. 2016: 70–77; Ji, Li-Jun et al. 2004: 57–65.
104   See Introduction, page 12.
105   Although the term used, ἀναγιγνώσκω, can also mean "recite from memory."
106   Gagné 2013: 297–316.
107   We hear about books available for sale in the market in a fragment of the comic play-
        wright Eupolis (fr. 327 K-A), and Plato has Socrates say that Anaxagoras's writings
        are for sale for one drachma (*Apology* 26D). See Reynolds and Wilson 2013: 1–43.
108   Pseudo-Plutarch, *Lives of the Ten Orators* 841F.
109   Such as the polychrome ivory statuette of a Roman actor for the first century CE
        in the Petit Palais, Paris. https://commons.wikimedia.org/wiki/File:Statuette_actor_
        Petit_Palais_ADUT00192.jpg#filelinks. Accessed November 11, 2016.

Bibliography

Anderson, C.A. and Dix, K.T. 2014. "Λάβε τὸ βυβλίον: Orality and literacy in Aristophanes" in
    Scodel, R. (Ed.), *Between orality and literacy: Communication and adaptation in antiquity:
    Orality and literacy in the ancient world* (Vol. 10). Leiden, the Netherlands: Brill: 77–86.
Bakola, E. 2010. *Cratinus and the Art of Comedy*. Oxford: Oxford University Press.
Benson, P.J. and Perrett, D.I. 1994. "Visual processing of facial distinctiveness" in *Percep-
    tion* 23.1: 75–93.
Blackburne, L.K., Eddy, M.D., Kalra, P., Yee, D., Sinha, P. and Gabrieli, J.D. 2014. "Neural
    correlates of letter reversal in children and adults" in *PloS One* 9.5: e98386.
Blais, C., Jack, R.E., Scheepers, C., Fiset, D. and Caldara, R. 2008. "Culture shapes how
    we look at faces" in *PloS One* 3.8: e3022.
Blundell, S., Cairns, D. and Rabinowitz, N. 2013. *Vision and viewing in ancient Greece*.
    Lubbock, TX: Texas Tech University Press.
Brennan, S.E. 1985. "Caricature generator: The dynamic exaggeration of faces by com-
    puter" in *Leonardo* 18.3: 170–178.
Brockett, O.G., Ball, R.J., Fleming, J. and Carlson, A. 2016. *The essential theatre*. Boston,
    MA: Cengage Learning.
Budelmann, F. and Easterling, P. 2010. "Reading minds in Greek tragedy" in *Greece &
    Rome* 57.2: 289–303.
Calame, C. 1986. "Facing otherness: The tragic mask in ancient Greece" in *History of
    Religions* 26.2: 125–142.
Calame, C. and Burk, P.M. 2005. *Masks of authority: Fiction and pragmatics in ancient
    Greek poetics*. Burk, P.M. (Tr.). Ithaca, NY: Cornell University Press.
Callahan, W.A. 1989. "Discourse and perspective in Daoism: A linguistic interpretation of
    ziran" in *Philosophy East and West* 39.2: 171–189.
Chambon, V., Vernet, M., Martin, F., Baudouin, J.Y., Tiberghien, G. and Franck, N. 2008.
    "Visual pattern recognition: What makes faces so special?" in Leeland, K.B. (Ed.), *Face
    recognition: New research*. New York: Nova Publishers: 71–90.

Chaniotis, A. 1997. "Theatricality beyond the theater: Staging public life in the Hellenistic world" in *Pallas* 47: 219–259.

Clark, A. 2015. *Surfing uncertainty: Prediction, action, and the embodied mind*. Oxford: Oxford University Press.

Csapo, E. 1997. "Riding the phallus for Dionysus: Iconology, ritual, and gender-role de/construction" in *Phoenix* 51: 253–295.

Csapo, E. and Slater, W.J. 1995. *The context of ancient drama*. Ann Arbor, MI: University of Michigan Press.

Dehaene, S. 2013. "Inside the letterbox: How literacy transforms the human brain" in *Cerebrum* 7: 1–16.

Dodds, E.R. 1944 (1960). *Euripides Bacchae*. Oxford: Oxford University Press.

Easterling, P.E. 1973. "Presentation of character in Aeschylus" in *Greece & Rome* 20: 3–19.

Easterling, P.E. 1977. "Character in Sophocles" in *Greece & Rome* 24: 121–129.

Easterling, P.E. 1990. "Constructing character in Greek tragedy" in Pelling, C. (Ed.), *Characterization and individuality in Greek literature*. Oxford: Oxford University Press: 83–99.

Easterling, P.E. 1997. *The Cambridge companion to Greek tragedy*. Cambridge: Cambridge University Press.

Ekman, P. 1999. "Facial expressions" in *Handbook of Cognition and Emotion* 16: 301–320.

Ekman, P. 2007. *Emotions revealed: Recognizing faces and feelings to improve communication and emotional life*. New York: Palgrave Macmillan.

Ekman, P. and Cordaro, D. 2011. "What is meant by calling emotions basic" in *Emotion Review* 3.4: 364–379.

Elam, K. 2002. *The semiotics of theatre and drama*. Hove, UK: Psychology Press.

Elman, P. 1999. "Basic emotions" in Dalgleish, T. and Power, M.J. (Eds.), *Handbook of cognition and emotion*. Chichester, UK: Wiley: 45–60.

Elsner, J. 2006. "Reflections on the 'Greek Revolution' in art: From changes in viewing to the transformation of subjectivity" in Goldhill, S. and Osborne, R. (Eds.), *Rethinking revolutions through ancient Greece*. Cambridge: Cambridge University Press: 68–95.

Emigh, J. 2003. "Playing with the past: Visitation and illusion in the mask theatre of Bali" in Schechter, J. (Ed.), *Popular theatre: A sourcebook*. London/New York: Routledge: 107–128.

Engelmann, J.B. and Pogosyan, M. 2012. "Emotion perception across cultures: The role of cognitive mechanisms" in *Frontiers in Psychology* 4: 118–118.

Farenga, V. 2014. "Open and speak your mind" in Wohl, V. (Ed.), *Probabilities, hypotheticals, and counterfactuals in ancient Greek thought*. Cambridge: Cambridge University Press: 84–100.

Foley, H.P. 1980. "The masque of Dionysus" in *Transactions of the American Philological Association* 110: 107–133.

Gagné, R. 2013. "Dancing letters: The alphabetic tragedy of Kallias" in Gagné, R. and Hopman, M.G. (Eds.), *Choral mediations in Greek tragedy*. Cambridge: Cambridge University Press: 297–316.

Gelder, B. de. 2009. "Why bodies? Twelve reasons for including bodily expressions in affective neuroscience" in *Philosophical Transactions of the Royal Society B: Biological Sciences* 364.1535: 3475–3484.

Gill, C. 1986. "The question of character and personality in Greek tragedy" in *Poetics Today* 7: 251–273.

Goffman, E. 1974. *Frame analysis: An essay on the organization of experience.* Cambridge, MA: Harvard University Press.

Goldhill, S. 1990. "Character and action, representation and reading: Greek tragedy and its critics" in Pelling, C. (Ed.), *Characterization and individuality in Greek literature.* Oxford: Oxford University Press: 101–127.

Gombrich, E.H. 1973. "The mask and the face" in Gombrich, E.H., Hochberg, J. and Black, M. (Eds.), *Art, perception, and reality.* Baltimore, MD: Johns Hopkins University Press: 1–46.

Gould, J. 1978. "Dramatic character and 'Human Intelligibility' in Greek tragedy" in *Proceedings of the Classical Philological Society* 24: 43–67.

Gregory, R.L. 1970. *The intelligent eye.* New York: McGraw-Hill.

Gregory, R.L. 1997. "Knowledge in perception and illusion" in *Philosophical Transactions of the Royal Society B: Biological Sciences* 352.1358: 1121–1127.

Griffith, R. 1998. "Corporality in the ancient Greek theatre" in *Phoenix* 52: 230–256.

Grisolia, R. and Rispoli, G.M. (Ed.). 2003. *Il personaggio e la maschera.* Atti del Convegno Internazionale di Studi, Quaderni del Centro Studi Magna Grecia 3. Naples, Italy: Bossura.

Halberstadt, J., Winkielman, P., Niedenthal, P.M. and Dalle, N. 2009. "Emotional conception how embodied emotion concepts guide perception and facial action" in *Psychological Science* 20.10: 1254–1261.

Hall, E. 2006. *The theatrical cast of Athens: Interactions between ancient Greek drama and society.* Oxford: Oxford University Press.

Hall, P. 2012. *Exposed by the mask: Form and language in drama.* London: Oberon Books.

Halliwell, S. 1993. "The function and aesthetics of the Greek tragic mask" in Slater, N.W. and Zimmermann, B. (Eds.), *Intertextualitat in der griechisch-römischen Komödie.* Stuttgart, Germany: Meltzlerschen & Poeschel: 195–211.

Halliwell, S. 2009. *The aesthetics of mimesis: Ancient texts and modern problems.* Princeton, NJ: Princeton University Press: 122–124.

Han, S. and Ma, Y. 2014. "Cultural differences in human brain activity: A quantitative meta-analysis" in *NeuroImage* 99: 293–300.

Hietanen, J.K. and Astikainen, P. 2013. "N170 response to facial expressions is modulated by the affective congruency between the emotional expression and preceding affective picture" in *Biological Psychology* 92.2: 114–124.

Huang, L. and Galinsky, A.D. 2011. "Mind – body dissonance conflict between the senses expands the mind's horizons" in *Social Psychological and Personality Science* 2.4: 351–359.

Imai, M., Kanero, J. and Masuda, T. 2016. "The relation between language, culture, and thought" in *Current Opinion in Psychology* 8: 70–77.

Jack, R.E. 2013. "Culture and facial expressions of emotion" in *Visual Cognition* 21.9–10: 1248–1286.

Jack, R.E., Garrod, O.G. and Schyns, P.G. 2014. "Dynamic facial expressions of emotion transmit an evolving hierarchy of signals over time" in *Current Biology* 24.2: 187–192.

Ji, L.J. and Yap, S. 2016. "Culture and cognition" in *Current Opinion in Psychology* 8: 105–111.

Ji, L.J., Zhang, Z. and Nisbett, R.E. 2004. "Is it culture or is it language? Examination of language effects in cross-cultural research on categorization" in *Journal of Personality and Social Psychology* 87.1: 57–65.

Jones, J. 1980. *On Aristotle and Greek tragedy.* Palo Alto, CA: Stanford University Press.

Kawai, N., Miyata, H., Nishimura, R. and Okanoya, K. 2013. "Shadows alter facial expressions of Noh masks" in *PloS One* 8.8: e71389.

Kirsh, D. 2009. "Projections, problem space, and anchoring" in Taatgen, N.A. and van Rijn, H. (Eds.), *Proceedings of the 31st annual conference of the Cognitive Science Society*. Austin, TX: Cognitive Science Academy: 2310–2315.

Kleffner, D.A. and Ramachandran, V.S. 1992. "On the perception of shape from shading" in *Perception & Psychophysics* 52.1: 18–36.

Konstan, D. 2006. *The emotions of the ancient Greeks: Studies in Aristotle and classical literature* (Vol. 5). Toronto: University of Toronto Press.

Korshak, Y. 1987. *Frontal faces in Attic vase painting of the Archaic period*. Chicago Ridge, IL: Ares Pub.

Lada-Richards, I. 2002. "The subjectivity of Greek performance" in Easterling, P. and Hall, E. (Eds.), *Greek and Roman actors: Aspects of an ancient profession*. Cambridge: Cambridge University Press: 395–418.

Lada-Richards, I. 2005. "Greek tragedy and western perceptions of actors and acting" in Gregory, J. (Eds.), *A companion to Greek tragedy*. Hoboken, NY: Blackwell: 459–471.

Lee, Y.J., Greene, H.H., Tsai, C.W. and Chou, Y.J. 2015. "Differences in sequential eye movement behavior between Taiwanese and American viewers" in *Frontiers in Psychology* 7: 697. http://doi.org/10.3389/fpsyg.2016.00697.

Lindberg, D.C. 1976. *Theories of vision form Al-Kindi to Kepler*. Chicago: University of Chicago Press.

Liu, Z., Zheng, X.S., Wu, M., Dong, R. and Peng, K. 2013. "Culture influence on aesthetic perception of Chinese and western paintings: Evidence from eye movement patterns" in *Proceedings of the 6th International Symposium on Visual Information Communication and Interaction*. New York: ACM: 72–78.

Livingstone, M. and Hubel, D.H. 2002. *Vision and art: The biology of seeing* (Vol. 2). New York: Harry N. Abrams.

Lloyd, G. 2007. *Cognitive variations: Reflections on the unity and diversity of the human mind*. Oxford: Oxford University Press.

Lyons, M.J., Campbell, R., Plante, A., Coleman, M., Kamachi, M. and Akamatsu, S. 2000. "The Noh mask effect: Vertical viewpoint dependence of facial expression perception" in *Proceedings of the Royal Society of London B: Biological Sciences* 267.1459: 2239–2245.

McCart, G. 2007. "Masks in Greek and Roman theatre", in McDonald, M. and Walton, M. (Eds.), *The Cambridge companion to Greek and Roman theatre*. Cambridge: Cambridge University Press: 247–267.

Marshall, C.W. 1999. "Some fifth-century masking conventions" in *Greece and Rome (Second Series)* 46.2: 188–202.

Masuda, T. and Nisbett, R.E. 2001. "Attending holistically versus analytically: Comparing the context sensitivity of Japanese and Americans" in *Journal of Personality and Social Psychology* 81.5: 922–934.

Meineck, P. 2011. "The neuroscience of the tragic mask" *Arion: A Journal of Humanities and the Classics* 19.1: 113–158.

Merleau-Ponty, M. and Toadvine, T. 2007. *The Merleau-Ponty reader*. Evanston, IL: Northwestern University Press.

Minoshita, S., Satoh, S., Morita, N., Tagawa, A. and Kikuchi, T. 1999. "The Noh mask test for analysis of recognition of facial expression" in *Psychiatry and Clinical Neurosciences* 53.1: 83–89.

Missiou, A. 2011. *Literacy and democracy in fifth-century Athens*. Cambridge: Cambridge University Press.

Mobbs, D., Weiskopf, N., Lau, H.C., Featherstone, E., Dolan, R.J. and Frith, C.D. 2006. "The Kuleshov effect: The influence of contextual framing on emotional attributions" in *Social Cognitive and Affective Neuroscience* 1.2: 95–106.

Mori, M. 1970. "The uncanny valley" in *Energy* 7.4: 33–35 (in Japanese).

Mori, M., MacDorman, K.F. and Kageki, N. 2012. "The uncanny valley [from the field]" in *IEEE Robotics & Automation Magazine* 19.2: 98–100.

Nielsen, I. 2002. *Cultic theatres and ritual drama: A study in regional development and religious interchange between East and West in antiquity.* Aarhus, Denmark: Aarhus University Press.

Nisbett, R.E. 2004. *The geography of thought: How Asians and Westerners think differently . . . and why.* New York: Simon and Schuster.

Nisbett, R.E. and Masuda, T. 2003. "Culture and point of view" in *Proceedings of the National Academy of Sciences* 100.19: 11163–11170.

Or, C.C.F., Peterson, M.F. and Eckstein, M.P. 2015. "Initial eye movements during face identification are optimal and similar across cultures" in *Journal of Vision* 15.13: 1–25.

Panksepp, J. and Watt, D. 2011. "What is basic about basic emotions? Lasting lessons from affective neuroscience" in *Emotion Review* 3.4: 387–396.

Pappas, A. 2001. "The aesthetics of ancient Greek writing" in Dalbello, M. and Shaw, M.L. (Eds.), *Visible writings: Cultures, forms, readings.* New Brunswick, NJ: Rutgers University Press: 37–54.

Parvizi, J., Jacques, C., Foster, B.L., Withoft, N., Rangarajan, V., Weiner, K.S. and Grill-Spector, K. 2012. "Electrical stimulation of human fusiform face-selective regions distorts face perception" in *The Journal of Neuroscience* 32.43: 14915–14920.

Pickard-Cambridge, A.W., Gould, J. and Lewis, D.M. 1968. *The dramatic festivals of Athens.* Oxford: Oxford University Press.

Plamper, J. 2015. *The history of emotions: An introduction.* Oxford: Oxford University Press.

Raschle, N.M., Zuk, J. and Gaab, N. 2012. "Functional characteristics of developmental dyslexia in left-hemispheric posterior brain regions predate reading onset" in *Proceedings of the National Academy of Sciences* 109.6: 2156–2161.

Reynolds, L.D. and Wilson, N.G. 2013. *Scribes and scholars: A guide to the transmission of Greek and Latin literature.* Oxford: Oxford University Press.

Rivolta, D. 2014. "Cognitive and neural aspects of face processing" in Rivolta, D. (Ed.), *Prosopagnosia.* Berlin/Heidelberg: Springer: 19–40.

Roselli, D.K. 2011. *Theater of the people: Spectators and society in ancient Athens.* Austin, TX: University of Texas Press.

Rossion, B., Hanseeuw, B. and Dricot, L. 2012. "Defining face perception areas in the human brain: A large-scale factorial fMRI face localizer analysis" in *Brain and Cognition* 79.2: 138–157.

Rudolph, K. 2015. "Sight and the presocratics: Approaches to visual perception in early Greek philosophy" in Squire, M. (Ed.), *Sight and the ancient senses.* London: Routledge: 36–53.

Sacks, O. 2010. *Musicophilia: Tales of music and the brain.* Toronto: Vintage Canada.

Seidensticker, B. 2008. "Character and characterization in Greek tragedy" in Revermann, M. and Wilson, P. (Eds.), *Performance, iconography, reception: Studies in honour of Oliver Taplin.* Oxford: Oxford University Press: 333–346.

Simpson, E.A., Murray, L., Paukner, A. and Ferrari, P.F. 2014. "The mirror neuron system as revealed through neonatal imitation: presence from birth, predictive power, and evidence of plasticity" in *Philosophical Transactions of the Royal Society B* 369.1644: 1471–2970.

Skoyles, J.R. 2008. "Why our brains cherish humanity: Mirror neurons and colamus humanitatem" in *Avances en Psicología Latinoamericana* 26.1: 99–111.

Sourvinou-Inwood, C. 2003. *Tragedy and Athenian religion*. Lanham, MD: Lexington Books.

Taplin, O. 1977. *The stagecraft of Aeschylus: The dramatic use of exits and entrances in Greek tragedy*. Oxford: Oxford University Press.

Tatler, B.W., Wade, N.J., Kwan, H., Findlay, J.M. and Velichkovsky, B.M. 2010. "Yarbus, eye movements, and vision" in *i-Perception* 1.1: 7–27.

Thomas, R. 1995. "Written in stone? Liberty, equality, orality and the codification of law" in *Bulletin of the Institute of Classical Studies* 40.1: 59–74.

Thomas, R. 2001. "Literacy in ancient Greece: Functional literacy, oral education, and the development of a literate environment" in Olson, D.R. and Torrance, N. (Eds.), *The making of literate societies*. Hoboken, NJ: Blackwell: 68–81.

Thumiger, C. 2007. *Hidden paths: Self and characterization in Greek tragedy: Euripides' Bacchae*. BICS Supplements: Vol. 99. London: Institute of Classical Studies.

Torrance, I. 2013. *Metapoetry in Euripides*. Oxford: Oxford University Press.

Van den Stock, J., Vandenbulcke, M., Sinke, C.B., Goebel, R. and de Gelder, B. 2014. "How affective information from faces and scenes interacts in the brain" in *Social Cognitive and Affective Neuroscience* 9.10:1481–1488.

Ventura, P., Fernandes, T., Cohen, L., Morais, J., Kolinsky, R. and Dehaene, S. 2013. "Literacy acquisition reduces the influence of automatic holistic processing of faces and houses" in *Neuroscience Letters* 554: 105–109.

Vovolis, T. and Zamboulakis, G. 2007. "The acoustical mask of Greek tragedy" in *Didaskalia* 7.1: 1–7.

Vuilleumier, P., Armony, J.L., Driver, J. and Dolan, R.J. 2003. "Distinct spatial frequency sensitivities for processing faces and emotional expressions" in *Nature Neuroscience* 6.6: 624–631.

Wade, N.J. and Wade, N. 2000. *A natural history of vision*. Cambridge, MA: MIT Press.

West, M.L. 1989. "The early chronology of Attic tragedy" in *The Classical Quarterly (New Series)* 39.1: 251–254.

Wieser, M.J., Gerdes, A.B., Büngel, I., Schwarz, K.A., Mühlberger, A. and Pauli, P. 2014. "Not so harmless anymore: How context impacts the perception and electrocortical processing of neutral faces" in *NeuroImage* 92: 74–82.

Wiggett, A.J. and Downing, P.E. 2008. "The face network: Overextended? (Comment on: 'Let's face it: It's a cortical network' by Alumit Ishai)" in *Neuroimage* 40.2: 420–422.

Wiles, D. 2007. *Mask and performance in Greek tragedy: From ancient festival to modern experimentation*. Cambridge: Cambridge University Press.

Wolf, M. 2012. "How the brain adapted itself to read: The first writing systems" in Blum, S. (Ed.), *Making sense of language*. Oxford: Oxford University Press: 24–50.

Yarbus, A.L. 1967. *Eye movement and vision*. Haigh, B. (tr.). New York: Plenum Press.

4 *Dianoia*

Intention in action

In *Poetics*, Aristotle places *dianoia* ("thought" or "reasoning") third after *muthos* and *ethos* in his list of the elements of tragedy and describes it as "being able to say what is possible and appropriate." For him *dianoia* is "that with which people demonstrate that something is or is not" (*Poetics* 1450b10–12). Aristotle associates *dianoia* primarily with speech acts, but here I want to take his description of how people demonstrate thought or intention into the realm of movement and consider how dance, gestures, and the other physical actions of the performers acted on the spectators who watched them. Nietzsche's wrote that the nature of the Greek mind was made manifest in dance, "since in dance the maximum power is only potentially present, betraying itself in the suppleness and opulence of movement."[1] Nietzsche's Dionysian man reaches his zenith in performance when he dances and becomes a "member of a higher community . . . on the brink of taking flight as he dances."[2] Nietzsche was right in proposing that Greek tragedy had its origins in the dances of the chorus and that movement remained an essential element of ancient drama.

In antiquity the close links between movement and drama were well established, and the legendary founding fathers of the theatre – Thespis, Pratinas, and Phynichus – were known as "dancing poets" (Athenaeus 1.39). Likewise, in Sophocles' *Oedipus Tyrannus* (895–896), when the chorus are faced with a debilitating religious crisis, they exclaim "why should I dance?" as if the very act of stopping dancing can bring the entire enterprise of drama to a complete halt.[3] With choral dances, synchronized movements, heightened gestures, powerful entrances and exits, and deliberate blocking (the movements of actors in a performance space), Greek drama was far from the monolithic, fixed tableaux, static pageant it has seemed to so many to be, but instead a dynamic and fluid theatre of almost constant movement.

This chapter explores dramatic movement from a number of interlinked cognitive perspectives to try to gain some understanding of its impact in performance and its contribution to the overall emotional effect of the plays. I am not concerned here with trying to describe Greek dance from a choreographic perspective or in establishing a typology of stage blocking and gestures. Instead, I will consider how the movement of the chorus and actors communicated motor-action-orientated kinesthetic empathy and sensory-mirroring to the audience and how the mask enhanced this process. I also explore the role of the so-called mirror neuron system in processing the movements of others and examine the value of neuroscience research in this area in helping to increase our understanding of the cognitive perceptual and affective properties of watching movement and dance. I then discuss an important facet of movement to the performative fabric of Greek drama – the procession – and suggest that it was not only on the periphery of the dramatic festival but also a central element in the development and form that Greek theatre took in the fifth century. Finally, I will briefly examine gestures, not as part of a system of semiotic symbols, but as fully embodied distributive communicators of intention, action, and emotion. To begin, movement will be placed within the predictive generative model to understand how thinking about movement – considering our future moves and understanding the movements of others – is essential to human cognition.

Perception and action

Imagine you arrive home late at night; as you get out of your car and walk toward your porch, you notice that the light is off and think "that's strange." Then you hear the faint noise of movement coming from the general direction of the porch, you freeze in place and rapidly turn your head in the direction of the sound, narrowing your foveal (focused) gaze, trying to peer through the dark to see what could have made that sound. Your first prediction is that it is the neighbor's cat who often visits your porch, but the lack of light and the quality of the noise combined with your racing heart rate, tensed body, and heightened sensory systems all strongly suggest that this could well be a prediction error. You search for more sensory information and move tentatively toward the porch in the hope of visual or aural clarification. As you do, you catch sight of something small, glinting in what little light there is – it is moving a little, up and down. Now your emotional systems come into play; you feel fear and experience its associated embodied responses – goosebumps, raised temperature, increased heart and breathing rate, and the flow of adrenaline – as you pay acute attention to the glinting object to refine your earlier prediction. This is not a cat, but perhaps it is a knife or a gun moving in the darkness – this could be an intruder, you could be in danger – then another sudden noise sends you reeling away and you turn and run back to the car, only to see the neighbor's cat with a small, shiny toy in its mouth dash across the lawn.

Clark uses such an example to demonstrate the importance of action to the generative predictive model of perception. It is here that action plays a key cognitive role as the best way to sample the environment to yield more reliable information and send it up the sensorial chain for evaluation. As new information is processed to help reduce uncertainty, our attention is heightened, our emotional state is primed for fight or flight, every part of our body is tense and ready to receive the commands to move or stay completely still to maximize perception. What is happening here as Clark puts it is that

> perception and action form a virtuous, self-fueling, circle in which action [such as turning the head, walking nearer, keeping still, evaluating the movement of whatever is on the porch, running] serves up reliable signals that recruit percepts that both determine and become confirmed or disconfirmed in action.[4]

This perceptive sequence is an embodied, distributed program, and our neuronal systems are not just trying to predict the incoming sensory signals, "they are constantly *bringing about* the sensory stream by causing bodily movements that selectively harvest new sensory stimulations. Perception and action are thus locked in a kind of endless circular embrace."[5] Aristotle gets at something similar in *Poetics* (1448a27–29) where he describes drama as "representing men in action and doing things" and "that some claimed that the word *drama* was derived from *drõntas*" ("doing"). Aristotle's famous definition of tragedy at *Poetics* 1449b25–30 distinguishes drama from epic in that the parts of the play are represented by people acting and not narration. For him, tragedy is *mimesis praxeos* – the representation of action as the result of some type of movement, whether transactional, physical, or predictive.

In the introduction to this book I proposed that theatre is a kind of mimetic mind in that it works because it re-creates and represents the cognitive processes of its audiences. The way in which dramatists create theatrical worlds on stage is an enacted replica of the generative predictive model, where our attention is focused on visual or aural information that builds a cognitive picture of a scene. Like the ambiguous mask, less is very often more in the surrogate environment of the theatre, as our evaluative systems need only the merest of information to start to actively generate the inferences we need to make sense of what we are experiencing. We see this operating in the surviving texts, such as the opening of Sophocles' *Philoctetes* where the simple fifth-century performing space consisting of *orchestra*, small stage, *skene*, and *eisodoi* are transformed into the beach and cliffs of Lemnos by the words of the characters and their related movements. Sophocles uses the same type of spatial designations to create a sacred grove in *Oedipus at Colonus* or the Greek camp at Troy in *Ajax*. Thus merest of information connected to movement can turn a small wooden scene building into the mythical house of Atreus by the Watchman in Aeschylus's *Agamemnon*. Even the most fantastical realms could be imagined on stage by the cognitive scaffolding offered by words with movements. For example, the Hoopoe's nest in Aristophanes *Birds* is created by the ridiculous vision of two men wandering around unfamiliar territory being led by a jackdaw. The castle in the sky that is the illusory *Cloudcuckooland* is created by having the actors gaze up into the extrapersonal sky, the ornithological costumes and movements of the chorus, and the actors making puns on how the wings of the birds are blocking the wing-entrances of the performance space (*Birds* 296-297). In this way, gaze direction, movement, visuality, and words all come together to conflate the physical surrogate theatre space with the fantastical world of the play.

Ancient Greek drama relied on *phantasia* and the predictive systems of its audience to comprehend the narratives that were played out in the minimal performance space at the theatre of Dionysos: minimal in the modern sense of non-representational or realistic, meaning without specific scenic elements, multiple costume changes, lighting or sound effects, elaborate stage furniture, or multiple stage properties. Without the kind of localizing theatrical elements we have come to associate with the theatre today, movement becomes vitally important to establishing a sense of place, defining its boundaries, and then communicating the mood and emotional tone of the scene to be played there. Before we delve into the interactive properties and products of movement and perception let us focus on how movement contributes to the establishment of narrative and informs the intentionality, or *dianoia*, of a certain scene. My example is the opening of Sophocles' *Ajax*.

As we saw in Chapter Two, the Athenian audience's experience of the performative elements of the festival of Dionysos were not limited to the confines of the theatre space. Before taking a seat to watch *Ajax*, they would have participated in a great procession, experienced the large dithyrambic choruses a day or two before, and possibly already watched several other dramas. We also know that the view of the city, landscape, and sky was always in the spectator's optic range and that there was constant sacrificial activity going on in the sanctuary immediately below the theatre. The Athenian was also surrounded by fellow countrymen and foreign ambassadors and would have looked down on a performing area with no stage curtain or proscenium frame, but rather a large open *orchestra* with a wooden scene building with one door and an upper level and perhaps a small raised stage before it placed on the edge of the *orchestra* opposite the *theatron*. There was probably no painted scenery and nothing to indicate time or place.[6] The Athenian may have known the theme and basic concept of the play beforehand and that there would be a masked chorus and three actors, at least two of whom would play several roles in different masks, as well as a musician playing the twin-reed pipe – the *aulos*. The Athenian would certainly be expecting music, lyrics, and dancing. Everything else, all other relevant information to create the world of the play, had to be communicated by spoken and sung narrative, gestures, and movement.

Ajax begins with a strange scene that was likely to have immediately focused the attention of the audience on a solitary figure who comes tentatively into the performance space. This figure is looking down, searching, and probably acting hyper-aware of his surroundings. The mask denoted a middle-aged male figure, and the costume could have suggested military garb, although we have no evidence for what he was wearing. The actor's movements here start to build the scene: he is searching for something, he is anxious, and it is not safe. The tension created by these movements would have been palpable, especially to an audience familiar with the tension of the battlefield and the armed camp. Then, another figure emerges in the performance space, or perhaps the actor was already in place, we cannot be certain, but the costume and mask quickly indicate that this is a representation of a goddess. The costume may have also included recognizable elements from Greek iconography such as a helmet, aegis, Gorgon head, and spear indicating Athena.[7] One might therefore assume that the goddess enters or was placed on the roof of the *skene*, although how this was achieved with any grace is hard to fathom (did the actor climb a ladder, or was he already in place and revealed somehow?). Athena may have entered from the *skene* door, which would certainly have provided the element of surprise. One other theory is that the actor is suspended on the *mechane* – the stage crane that was certainly used by Euripides to move gods from backstage to the *skene* or the *orchestra*.[8] We cannot know for sure. What the text does indicate is that this goddess figure instantly names Odysseus and describes how she sees him "hunting," then Odysseus says that he hears the voice of Athena, as he says he cannot see her, perhaps an indication that she is on the roof – in sight of the audience but not the actor – who, as we know from the way the mask operated in the space, would have had to face front, or at least no more than a three-quarter turn, whenever he spoke in order not to lose vocal audibility.

Odysseus tells Athena that he is following the tracks of the animals that have been dragged from the Greek camp and that many more were slaughtered there. Athena reveals it was Ajax, his ally, who committed this atrocity, and he was actually trying to kill the Greek commanders, but his perception was tricked by the goddess, who made him believe that the animals were actually the Greek commanders he had come to kill. Odysseus is appalled at this information and is then informed that Ajax is inside the *skene*, which has now been identified as the hut of Ajax at Troy. Ajax is described as being in the grip of his delusional madness and torturing animals from the Greek camp that he has dragged inside. Athena, acting like a theatre director, tells Odysseus that she will display Ajax in the midst of his madness, so, like a *theoric* traveler, Odysseus can experience the "performance" and then return to the Greek camp and tell the other Greeks what he has seen. Odysseus is paralyzed by fear, because if she brings Ajax out of the *skene* he will surely attack him. But Athena reassures him and says she will turn Ajax's eyes away from landing on Odysseus's face/mask (*prósōpon*). It is here that Athena makes the most basic of theatrical compacts: Odysseus must stay still and quiet and just watch as she calls on Ajax to "enter" from the *skene* door. Now Odysseus becomes an invisible audience and Ajax the visible actor.

This scene provides us with an excellent metaphor for the basic conditions of drama, or what Aristotle identified as mimesis *in action*. I attempted to reconstruct what may have happened on stage during the opening scene of *Ajax* by using the surviving text and combining it with what we know of the staging conditions, environment, and culture of mid-fifth-century Athens. For this, the text alone is not enough, drama is not literature, it is meant to be enacted and enlivened by movement. Thus, Aristotle suggested that a playwright should "put the events before his eyes as much as he can," performing the actions of the play they are composing in order that narrative and action are seamlessly combined (*Poetics* 1455a22–25). Aristotle also tells us how an entire play will fail if movements and words are not properly coordinated and mentions a work by a fourth-century playwright, Carcinus, whose character, Amphiaraus, enters from a shrine. This must have jarred with the spoken narrative, and Aristotle says that this would be missed by anyone not seeing the play, presumably because the stage directions were not available to the reader, but very clear to the spectators, who were apparently very upset about it, and the play was regarded as a flop (*Poetics* 1455a27–29).

Reconstructing the movements of Greek drama is very difficult. We do have some evidence for the movements of actors, including textual references, images from vase painting and sculpture, and later handbooks for orators on gestures, but these cannot re-create how the actors moved in performance and only provide fleeting and static glimpses frozen in stone or text. What we can do, however, is hypothesize how stage movements affected the audience who watched them and contributed to the communication of emotions and empathy.

The action-orientated spectator

Although in *Ajax*, Athena instructs Odysseus to keep still, on a cognitive level the act of watching another's movements is far from a passive activity. There is now a good deal of evidence that shows how in processing the movements of others we are activating parts of our own mental systems involved in the production of movement. This is, in effect, an embodied feedback loop of action and a powerful component in the communication of drama from the actor to the spectator. This is borne out by times in which staged action and narrative information conflict and create a cognitive rift by causing a "stoppage" in the predictive processing chain that can't be resolved. These moments can abruptly pull audiences out of the kind of cognitive absorption needed to sustain an empathetic response. This might be when an actor obviously stumbles over a line or a piece of stage business seems too contrived or at odds with the overall style of the performance. An ancient example of this is provided by the unfortunate actor Hegelochus, who is remembered for mispronouncing a line from Euripides' *Orestes* (279) so that instead of hearing "once more the storm is past I see the calm," the audience actually heard "I see the weasel." At such a heightened and emotional moment in the play, where Orestes is being driven insane by the Furies and his sister, Electra, is desperately trying to comfort him, such an error would have ruined the entire scene, if not the whole performance.[9]

Aristotle advises that the kind of errors he attributed to Carcinus can be avoided by the playwright, who must "complete his plots with gestures" and when composing try to replicate the emotional state they want their characters to display (*Poetics* 1455a27–29). Janko has suggested that what Aristotle means is that "the poet should perform his drafts to see whether they evoke the right emotions."[10] Thus, even at the moment of the play's creation, movements and gestures are combined with text to effectively activate emotional responses.

Aristotle makes the same point in his advice to public speakers in *Rhetoric* (2. 8.13–15). He writes that people tend to empathize with those who most closely resemble them or the most recent events. How then to elicit empathy or pity (*eleos*) for events from the distant past? He advises that they should combine gestures, voice, costume, and acting to place the action "before the eyes" of the audience to elicit empathy in the here and now. Likewise, in Aristophanes we encounter parodies of Athenian dramatists dressing as women or in rags as they compose their works to "feel" the emotionality of the narratives they are creating and imbue their works with this kind of embodied affective information.[11]

In *Ajax*, the frozen Odysseus mediated the tension of the audience – fellow spectators who gazed on the frenetic movements of the crazed warrior. The actor playing the role of Ajax may well have been holding a whip, which we hear he has been using to torture the animals inside his tent. The spectator becomes vicariously involved when Ajax announces that he is torturing Odysseus, who he believes in bound up inside his tent. Every time that whip stung the air, the fictional Odysseus may have mimetically shuddered – but the audience may well have also flinched at the bite of the scourge. Thus, the body's autonomic responses to the probability of the whip causing pain affects the mental state of the audience. In this way Ajax's destructive crazed power is *felt* by the audience and heightens their feelings of anxiety, fear, anger, or revulsion. This would have stood in marked contrast to the next time Ajax is revealed in the play, seated catatonically on the *ekkyklema* surrounded by the butchered carcasses of the animals he had mutilated the night before – a powerful visual manifestation of the present state of Ajax's psyche. His lack of movement here is as palpable and tension filled as his animated performance was earlier in the play.

Athena's stage management of Odysseus comes to fruition at the resolution of the play. After Ajax commits suicide, the play develops into a gripping debate as to whether he should be buried or not. The sons of Atreus, the primary targets of Ajax's night attack, want to leave him unburied as a traitor, but Teucer, his half-brother, stands up to them and presses his familial rights of burial. This argument threatens the stability of the Greek forces until Odysseus arrives as a kind of surrogate *deus ex machina* to intervene and bring peace. What has enabled Odysseus to do this is that he has experienced the downfall of Ajax first hand as a spectator, but more than that, he has felt for himself the predicament that Ajax found himself in. In effect, by being the audience, Odysseus has learned empathy by experience.

> I pity him in his misery, though he is my enemy, because he is bound fast by a cruel affliction, not thinking of his fate, but my own; because I see that all of us who live are nothing but ghosts, or a fleeting shadow.
>
> Sophocles' *Ajax* 121–126

Here, Odysseus says he feels pity for Ajax – *epoiktirō*, a term that means to "feel for" and somewhat equitable with the modern term "empathy." This term is also found in Aeschylus's *Agamemnon* 1069, when the chorus of Argive elders say they feel for the captive Trojan girl, Cassandra; it is uttered several times about the wounded warrior Philoctetes in Sophocles' *Philoctetes* (318, 1071 and 1320); and at the end of *Oedipus Tyrannus* (1474), Oedipus hopes that Creon will "feel for him" by letting his children come out and move closer.[12] In *Ajax* Odysseus takes on an almost semi-divine role as a figure of arbitration because he has witnessed the distress of another for himself depicted in vivid live action, and as a result has come to empathize with them. Greek drama sought to do the same thing – to absorb the audience, to communicate emotionality, and to create empathy – and this could only be achieved by placing the characters, plots, and visual elements *in action*. This was conveyed not only by the narrative of song and spoken word but also by what was communicated by the movements of the actors, what has been called kinesthetic empathy. This was particularly important within the choral culture of ancient Greece where movement and dance were collective experiences and the boundary between spectator and performer was fluid and permeable. This was true of the theatre where watching movement shared powerful cognitive associations with the act of moving itself.

Kinesthetic emotional contagion

In seeking to describe the affective abilities of the masked Roman Pantomime dancer (to be discussed further later), Ruth Webb has explained how Pantomime emphasized "the powerful inseparable union of movement and emotion in the dancer's art that communicates directly from body to body and leaves the audience vulnerable to the same emotion they are drawn to copy with the same movements."[13] What Webb is describing is kinesthetic empathy, the theory that affective states can be communicated by observing dance, gestures, bodily postures, and the facial and bodily movements of others.[14] This transference of emotions from performer to observer has been called "emotional contagion," where the affective states of others are "mirrored" by subliminal micro-expressions of the recipient responding to the target, a theory that was first advanced by Hatfield, Cacioppo, and Rapson.[15]

There are, of course, complexities; some people seem to be more receptive to emotional contagion than others, and the context whereby emotional contagion occurs also matters, especially within aesthetic forms such as musical performances, film, and theatre.[16] Cultural mores also play an important part in this kind of non-verbal communication. For example, one recent study examined emotional contagion between teachers and students and the teacher's perception of their students based on their own self-perception of how they should behave in a classroom. This kind of reflective mirroring based on the student's bodily signals as perceived by the teacher was shown to have an impact on classroom environment, learning, and morale.[17]

Emotional contagion is also not limited to visual perception, and studies of differing emotional telephone interactions have shown that there can be voice-to-voice emotional contagion.[18] Furthermore, a controversial study involving nearly 700,000 Facebook users suggested that people were being influenced by the emotional tone of the Facebook postings they were exposed to.[19] While this research was controversial in that participants unknowingly took part in the study and it exposed the weakness of human research ethical codes for big data, it did indicate that emotional contagion is spread in a variety of ways, and not all of them are physical.[20] One of the drawbacks of the Facebook study was that the physical embodied responses of the participants were not measured. This is a kind of reflexive emotional contagion where the body mirrors, often at a micro level, the affective visual signs of others. Perhaps the body also reacts to text in this way, helping the reader to feel an emotion. Future studies might explore the difference, if any, in bodily responses to text versus images, both of which are available in abundance on social media.

If we place emotional contagion within the theory of predictive processing, it falls broadly into Theory of Mind cognitive models – the way in which we infer the intentions and beliefs of another person. This relies on exteroceptive (such as sight and sound), proprioceptive (movement), and interoceptive (visceral) mechanisms. Friston's free energy principle states that the brain must generate continuous predictions to minimize "free energy." To do this we can either alter our top-down predictions to match the incoming sensory information or we can control our visceral system (motor and autonomic) to experience sensations that comport to our predictions or "active inferences." Emotional contagion and kinesthetic empathy, where we autonomically mimic the emotional expressivity of others, is a vital part of this visceral active inference system.[21] For example, we can infer that a character on stage is feeling afraid by observing the physical actions that display fear. However, the interoceptive cognitive mechanisms may also operate by having us feel a pit in our stomachs or a slight shudder as we also respond viscerally to the mimetic elements that project a sense of fear. These might be the tone of the music, narrative information, the actor's gestures, perception of the emotion of the mask, a change in meter, the vocal quality of the actor, and the common reactions of the audience around us. As Ondobaka et al. have pointed out, "deep [visceral] generative models permit inferences at multiple levels . . . and may induce an interoceptive or emotional contagion and empathy – implying that an observer can also be sympathetic to another's desires and intentions."[22]

Active inference involves the motor planning functions of the human mind, but in exteroceptive and proprioceptive terms this kind of autonomic "mirroring" has been linked to kinesthetic empathy.[23] This involves the brain's action observation network and the response of the so-called mirror neuron system, which correspond with the areas of the brain that have been identified as being most associated with motor planning and action.[24] We will return to questions surrounding the mirror neuron system soon, but in the meantime it is important to know that recent brain lesion studies have identified that a neural network involving the inferior frontal gyrus and the inferior parietal lobe are needed for emotional recognition and the possibility of emotional contagion.[25] There may well also be another empathetic pathway that utilizes the ventromedial prefrontal cortex, temporoparietal junction, and the medial temporal lobe in the processes of self-reflection and autobiographical memory, regions that have been shown to be necessary for the evaluation of other's affective states. Although these neural systems appear to work independently, both can come into play in the recognition and processing of affective responses. This neural hypothesis is further supported by clinical practice in dance, music, and drama therapies, where therapists and participants use mirroring techniques to gain a better understanding of another's psychological situation, in order to inform and aid treatment.[26] The importance of kinesthetic empathy in the Greek theatre will soon come into view, but first we need to tackle the role of the mirror neuron system in this process and how we might use relevant brain imaging research to further understand the effect of movement in the ancient theatre.

Mitigating mirror neurons

It has been proposed that the mirror neuron system facilitates connections between the visual and motor cortexes, allowing humans to quickly learn behavior through both observation and kinesthetic understanding. This theory was first advanced by a team from the University of Parma in Italy, led by Giacomo Rizzolatti, in the early 1990s. They were conducting neural implant research on macaque monkeys, so electrodes were attached to pre-motor areas of their brains and their neural responses to picking up food items were recorded. When a researcher inadvertently picked up the food item that had been situated for the test, the monkeys had the same neural response as if they had picked it up themselves.[27] This has led to an enormous amount of research to determine whether humans possess this same kind of action-perception matching mechanisms or even an "empathy response,"[28] and as Cook et al. have recently pointed out, "there is now a substantial body of evidence suggesting that mirror neurons are indeed present in the human brain."[29] But what exactly does the proposed presence of the mirror neuron system mean in theatrical terms, and how is it engaged in the act of watching acting – or "doing" – to use Aristotle's definition of drama?

Corradini and Antonietti have gathered many pertinent studies on the mirror neuron system and empathy, concluding that "empirical evidence shows that mirror neurons play a major role in the construction of a basic kind of relationship with the other."[30] Their survey of recent research also includes work on people paired with a confederate who imitates their body movements and posture during a shared task. Both participants scored higher on an empathy scale. Such physical "mirroring" devices have long been deployed as a technique of movement therapy, where the therapist is seeking to gain an empathetic insight into the patient under observation, and as an acting technique to better understand a given character who must be embodied by the actor.[31] Furthermore, it has been demonstrated that if people are shown pain being inflicted on another's body part, the muscles in their own corresponding part will excite, such as with a move to withdraw at the sight of a needle approaching another's hand.[32] At the same time as the muscles are excited, so the neurons in the anterior cingulate cortex respond as if the person observing the pain (the needle penetrating the other hand) is actually experiencing it. All of us have felt this "flinch" from time to time when being particularly absorbed by a dramatic imitation of pain while watching theatre or a film, just as the crack of Ajax's scourge may have produced such an empathetic wince. It has also been shown that the same area of the brain (the insula) that activates to disgust will also activate when a person looks at a face displaying a disgusted expression, and the same neuronal system is activated whether people smell a bad odor themselves or look at the face of a person smelling a bad odor.[33]

There can now be little doubt that based on the current science available, the mirror neuron system plays an important role in this aspect of human interpersonal social cognition. What particular perceived signals provoke an empathetic response must surely be conditioned by cultural interactions, but studies with newborns suggest that some level of kinesthetic empathy might be innate.[34] Yet the mirror neuron system does not act alone and is probably not the unilateral empathy-generating system that some have made it out to be. Clark is probably right to place the mirror neuron system within the sensory arsenal available to the brain when processing incoming visual motion-based information.[35] For example, although we might recognize the gesture of a raised arm, without additional contextual information and the kind of fluid error correction provided by the generative prediction model, we would not know, using the mirror neuron system alone, if the person making the gesture is waving hello or swatting a fly.[36] This is what is known as the inverse problem: though we may broadly recognize a gesture and even "feel it" by way of the empathetic response of our own mirror neuron system, we cannot by kinesthetic empathy alone know the gesture's action intentionality. As Kilner and Friston have put it, "the predictive coding account of the MNS specifies a precise role for the MNS in our ability to infer intentions and formalises the underlying computations. It also connects generative models that are inverted during perceptual inference with forward models that finesse motor control."[37]

With this in mind, we can place the human mirror neuron system within a more complex multi-directional (bottom-up/top-down) predictive cognitive system while still being able to benefit from a good deal of neuroscience research that has shown the human mirror neuron system in operation in experimental situations. In fact, the mirror neuron system supports the theory of predictive processing by engaging our neural mechanisms for kinesthetic empathy in the active process of prediction. The Parma monkeys learned the action of taking the food in the experiment by their own embodied experience; when they then saw the scientist inadvertently do the same thing, their predictive system sent the bottom-up sensory signals. These were refined by the mirror neuron system, which was also responding to the top-down evaluation based on working memory of prior experience. The human mirror neuron system may well have developed as one part of a neural system to help us understand the predicaments of others in order to enhance mutual survivability and group cohesion, and it may also play its part in helping to generate Theory of Mind – the ability to perceive another human's state of being and provide important sensory and embodied information for mutual interactive engagement.[38] The entire enterprise of theatre is one of active inference and Theory of Mind: without the ability to understand the intentions and motivations of others and the effect they might have, there can be no drama. Aristotle was surely correct to surmise that "tragedy is a representation not of people as such, but of actions and life" (*Poetics* 1450b17).

Long before the furor generated by the announcement of mirror neurons, people's abilities to recognize movement in others had been investigated. In 1973, Johansson created a series of films of what at first resemble random tiny dots.[39] In actuality, these are light diodes attached to the joints of a human figure that are impossible to discern until the figures move. Then it becomes perfectly clear that a human is being displayed engaged in different forms of movement such as walking or kicking a ball. What Johansson found is that from this most basic of information, humans can very quickly identify people known to them by the way they move and can even recognize themselves, which is all the more remarkable considering that most people do not watch themselves move. This is a characteristic of proprioception, which is the sense of the relative position of different parts of the body in relation to each other, or what might be termed the orientation of one's limbs in space. Friston, Daunizeau, et al. have proposed that proprioception is an essential component of the predictive motor system within the structure of the free energy principle, in that it corrects the error signals between where the body is situated in space and where it would be once the desired actions are carried out. They conclude:

> Perceptual learning and inference is necessary to induce prior expectations about the sensorium and action is engaged to resample the world to fulfill these expectations. This places perception and action in intimate relation [via proprioception] and accounts for the both with the same principle.[40]

In the predictive generative model, proprioception is as important for perception as exteroception (senses experienced "outside" of the body, such as hearing and sight) and interoception, which includes visceral responses such as hunger and thirst.[41] The masked performers of Greek drama had no peripheral vision and could not see their arms and feet or even each other for most of the time, even though the chorus members were required to dance and probably move together in unison. A heightened sense of proprioception and an acute spatial awareness was therefore an essential aspect of their stagecraft. For the spectators, this meant that they perceived movements and gestures in a precise and heightened manner, and although we cannot know for certain the dance steps, movements, and gestures of the performers, iconographic evidence and comparisons with other masked dramatic traditions would strongly suggest that this was the case. There is also a body of research that suggests that the mirror neuron system plays a role in the neural processing of proprioception.[42] This is a significant effect of the mask on the performer and can be very disconcerting for the actor who is not used to working this way. It must have also impacted the members of the chorus, whose own sense of self in space would have been mitigated by the mask and perhaps enhanced a sense of collectivity, which was kinesthetically communicated by the entire group.[43]

Kinesthetic empathy is generated by this kind of embodied motor-mimicry, a form of implicit bodily communication where we adopt similar postures, mannerisms, and bodily configurations to heighten social rapport and interpersonal communication.[44] Just as research has revealed that we move facial muscles empathetically, so studies have shown that the proprioceptive responses of some audience members can be stimulated when watching dance performances. These people can experience a kinesthetic sensation known as motor simulation, and their neural activity connected to the mirror neuron system increases significantly when the dance form performed is well known to the spectator.[45] This was demonstrated in 2005 by a team at University College London led by Patrick Haggard. In a controlled experiment, professional ballet dancers first watched ballet and then the Brazilian dance/martial art form known as capoeira, and then capoeira dancers watched capoeira followed by ballet.[46] The dancers watching their own dance form responded more strongly, suggesting the influence of motor expertise on action observation. Therefore, the neural mirror system integrates movements seen with movements known, and "the human brain understands actions by motor simulation."

Most of the spectators watching drama in classical Greece could be classified as "expert dancers," whether they were Athenians or from other Greek cities. Dance was an enormous part of Greek cultural identity, not to mention the equally synchronous movement activities of hoplite drill, rowing a trireme, riding in a cavalry formation, or being part of a procession. Of the Athenians, it might be safely said that almost everyone in attendance was, from an early age, highly familiar with dance as a cultural participatory activity. The City Dionysia involved 50 boys and 50 men from each of the 10 tribes of Attica competing in the dithyramb – a total of 1,000 performers recruited exclusively from the population of Athenian males. In addition, the tragedies and satyr dramas involved choruses of 12 to 15 and the comedies, choruses of 24, placing at least 108 chorus members in each City Dionysia (not to mention another 100 or so in the Lenaea). This means that the audience members almost certainly performed at one time themselves, if not at the Dionysia or another city festival, then at the very least in their *deme* or at family events.[47] The hymn sung and danced by the women of the chorus of Aristophanes' *Thesmophoriazusae* (947–1000) illustrates this strong connection between audience and dancer, as they invite the spectators to watch them form their circle dance. It is as if their appeal to join hands reached across the *orchestra* and out into the *theatron* to be felt by everybody. To *watch* dance was to *feel* dance: in this environment, spectatorship was also participation, which was a key element of *theoric* practice where the observer was as honored as the observed. Music and rhythm also enhanced this experience, as we shall see in the next chapter; it was as if the whole audience moved together.

> Come on and dance!
> Light feet forming the circle
> Join together, hand in hand
> Everyone feel the rhythm of the dance
> Quicker now, move those feet!
> Let everyone's eyes everywhere
> Watch the formation of our circle dance.
> Aristophanes, *Thesmophorizusae*,
> 953–958, tr. A. Hollmann

Whole body perception

In the last chapter I suggested that the Greek dramatic mask was capable of producing a heightened empathetic response in the spectator and that it operated as a kind of "supra-face" – its schematic and ambiguous features, manipulated correctly, were able to create a deeply personal kind of motor-action empathy in the observer by placing cognitive attention on the movements of the masked performer's body. The face is also an intricate moving part of the body. It is what neuroscientist John Skoyles has called a "motor exposure board" in that it contains hundreds of muscles capable of generating a number of easily identifiable macro-expressions and a much larger number of seemingly imperceptible "micro-expressions." When engaged in communication with another, the face is in a state of almost constant movement.[48] What's more is that studies have shown that when we watch a face, our own faces have a tendency to "mirror" the macro-expressions we perceive with our own micro-expressions – we embody the emotional states being communicated to us and can even start to feel those same emotions.[49] A way to experience this is to place a pencil lengthways between your teeth, which engages the facial muscles involved in smiling, and see if it has an effect on your mood.[50] It has also been observed that an infant will mirror the facial expressions of his or her caregiver.[51] As the child has never seen his or her own face, and therefore has no visual sense of how manipulating certain facial muscles produces a smile or a frown, it has been posited that this ability is innate and has also been connected to the function of the human mirror neuron system.[52]

The mask removed a key element of our emotional and empathetic contextual recognition mechanism – the moving face and the expressions it reveals; instead, ancient audience members had to rely on other contextual clues, including body movements, gesture, music, and language, all of which received increased cognitive attention as a result. Studies on patients with Moebius syndrome, a rare neurological condition where people cannot move their faces, has shown that such participants had trouble recognizing facial emotions.[53] Yet many of the same patients are able to perceive emotions when viewing the whole person, not just the face. This leads to the conclusion that our emotional processing systems are embodied and significantly enhanced by kinesthetic empathy, in that by mirroring micro-expressions we are able to more fully understand the emotional states being displayed to us by others. Because Moebius patients are unable to move their own faces, these studies also show that we receive kinesthetic emotional information from the body.[54]

New research on whole body perception can tell us a great deal about how the movements of the body with a mask were able to be so effective in the communication of kinesthetic empathy in performance.[55] Several important research findings stand out:[56] first, that the body is far better at communicating affective states over long distances than the face, which is only effective in close proximity. It has been pointed out that this shifts attention away from facial identification (gender, age, ethnicity, status etc.), which is not relevant for the rapid processing of the expressed affective state. Thus, bodies transfer action and behavioral responses more effectively than faces, and therefore there is a difference in the social communicative roles of the body and the face. When we focus on the face, we are prone to infer an individual's mental state, but attention to the body focuses on individual or group action. This is clearly relevant to large open-air Greek theatre spaces. Second, body expressions provide emotional context; studies have shown that people are often confused as to how to emotionally categorize a disembodied face displaying surprise.[57] When observing the same face in its corporeal context, emotion recognition becomes far higher. Third, although the cortical neural networks involved in face and body processing overlap within the fusiform gyrus, viewing whole body expressions utilizes a broader network, including the cortical and subcortical motor areas. Observing the whole body is more cognitively engaging than only looking at a face.[58]

We have already seen how Greek masks had emotionally ambiguous faces and how this helped them seem to change their expressions when they were moved. A 2007 whole body perception study showed participants angry and fearful faces and combined them with both congruent and incongruent angry and fearful body postures. The results showed that the bodily displayed emotions biased the identification of the facial expressions. When asked to rate how fearful or angry the faces were, the emotion displayed by the body posture proved most influential, and this influence was the greatest for the faces that were the most ambiguous.[59] Furthermore, de Gelder has suggested that neutral (or ambiguous) faces are actually processed in areas "more dominantly dedicated to body perception."[60] Performing in a mask activates the neural networks utilized for the rapid decoding of bodily displayed affective states. Therefore, the emotion states of the characters of Greek drama were not initially mentalized and contemplated but fully embodied by the masked performers and quickly and viscerally processed by the spectators. Far from the mask "distancing" the viewer, in tandem with movement and gesture it greatly enhanced emotional contagion and empathetic responses. This is reminiscent of Xenophon's account of Socrates visit to the artists in *Memorabilia* 3.109–115; his four key terms – *prósōpon* "face," *schêma* "form," *stasis* "stillness," and *kinêsis* "movement" – come together in the performer's mask and body to communicate affect and generate empathy.

To conclude this section and bring movement, mask, and action together, I take another example from Emigh's fascinating account of a Topeng performer in Bali in the 1970s and note how this training also emphasized the whole body in preparing to perform in this dance-drama form:

> The Topeng performer first learns dances by rote, his body being pushed and pulled into the proper shapes by his guru. Bit by bit, by imitation and direct manipulation of his limbs, he assimilates the dance vocabulary he will need to perform a range of masked characters. The masks themselves are not used at all at this stage of the training, yet to my surprise, when I was training with Kakul, he would repeatedly criticize my facial expressions as I grimaced while awkwardly trying to make my feet, arms and body work together in excruciatingly unfamiliar ways. The masks were not a disguise. If the face of the actor behind the mask did not register the character of the figure dancing, the body would move wrong, and the mask would be denied its "life".[61]

Sensorimotor gestures

A common trait in different masked performance dance/theatre traditions is a focus on gestures, finger, and even toe movements. We can observe this in Indian Kathakali, Japanese Noh, Balinese Topeng, and in depictions of masked actors on Greek vases (Figure 4.1). Theatre artists have implicitly known the cognitive

Figure 4.1 Attic red-figure krater from Spina, c. 450 BCE. (Museo Archeologico Ferrara. Valle Pega 173c). Boy with mask and masked maenad.

and communicate power of gestures. In the early 20th century, as Western theatre artists sought to explore what they perceived as the fundamental elements of earlier dramatic forms, the focus of gestures as significant communicators of action, emotion, and objectives became central to many theories of acting. For example, Brecht formed the concept of "gestus," which was an embodied sign system that sharpened the attitude of characters toward each other, both emotionally and socially, to enhance the audience's comprehension of the story, which he described as something that happens "between people" and "the complete fitting together of all the gestic incidents."[62] Likewise, the critic Walter Benjamin described the theatre of Franz Kafka as a place where "each gesture is an event, a drama unto itself" and "the gesture remains the decisive thing."[63] In the late 1960s, the Polish auteur Jerry Grotowski also articulated the importance of gesture in dramatic narrative, writing "by his controlled use of gesture the actor transforms the floor into a sea, a table into a confessional, a piece of iron into an animate partner."[64] Schechner describes Grotowski's "gesticulatory ideograms" as akin to the gestures of medieval European theater, Beijing opera, and ballet.[65] What many modern avant-garde theatre artists were trying to do was to create a theatre that emphasized the importance of the body and its gestures as at least as significant as the words the actors were speaking on stage.

Artaud lamented the primacy of cinema in the 1930s and felt that this new pop-ular art form was usurping the body in live performance. His focus on Balinese drama and older non-Western theatrical forms was in large part also an attempt to reinvigorate the importance of gestures. The surviving mask traditions of non-Western theatre that so captivated Artaud, Brecht, Craig, Yeats, Cocteau, Kafka, and many others has much in common with the performance elements of ancient Greek drama, and a common thread is the importance of gesture. This is appar-ent in the training of Kathakali performers in India and Noh actors in Japan, both forms that are heavily dependent on teaching the accurate and fluid embodiment of gestural and whole body sign systems. We also find the importance of gestures and bodily movements articulated by Bharata in the *Natyasastra*: "the behavior of people of various kinds; conveying that which by action and bodily movements is called drama."[66]

What is so fascinating about the avant-garde experiments of many of the thea-tre artists described earlier is that they were in part influenced by the work of the classicist Jane Harrison, who had come to see ancient drama as a movement-based expression of ritualized dance and gesture. Her revolutionary ideas pub-lished in several widely read works, such as *Prolegomena to the Study of Greek Ritual* (1903), *Themis, A Study of the Social Origins of Greek Religion* (1912), and *Ancient Art and Ritual* (1912), had a profound impact on dramatists such as W. B. Yeats, Granville Barker, TS Eliot, and the director Gordon Craig, all of whom cited her as an inspiration.[67] In America, Isadora Duncan, one of the preeminent founders of the modern dance movement, listed Harrison as a "reader" in a dance concert program in 1900, and Harrison was still inspiring avant-garde artists in the 1950s and 60s when the choreographer Merce Cunningham cited her as an important influence on his own work.[68] Harrison was influenced by the archaeo-logical work of Willhelm Dörpfeld who had excavated at the site of the Theatre of Dionysos in the late 1880s. While she may have been wrong to have so whole-heartedly embraced Dörpfeld's influential theories of a circular orchestra as the seminal element of the ancient stage, she was quite right in envisioning dance, movement, and gesture as fundamental to the origins and performance of Greek drama.

In every culture, people gesticulate when they speak, people who have been blind from birth use gestures, though they have never seen them, and both infants and primates use gestures alongside basic vocalizations. Gestures are a universal human communicative system, and although every culture has developed its own distinctive language of gesture, all use hands and arms to communicate.[69] Hutch-ins has described gestures as part of a distributed system of interpersonal commu-nication and as properties of the system of interaction, distinct from the cognitive properties of the individual who participates in the system.[70] But our gestural sys-tems are perhaps even more fundamental than only existing as a means of external communication: Malafouris points out how recent studies indicate that gesturing actually reduces cognitive load and enables the freeing up of other mental facul-ties such as memory retrieval. He also emphasizes the strong interdependence of hand and mind in the evolution of human cognition and points out that "the hand is not simply an instrument for manipulating an externally given objective world by carrying out the orders issued to it by the brain; it is instead one of the main perturbatory channels through which the world touches us."[71] This builds upon the seminal theories of embodied cognition of Lakoff and Johnson, who advanced the idea of gestures as metaphorical projections tied to our bodily sensorimotor systems that allow for the realization and communication of abstract concepts.[72]

The importance of gestures in interpersonal communication has been shown in a recent study where participants were asked to communicate to each other by either developing a brand-new vocalization system or with a new gestural system. Overwhelmingly, the gestural system was preferred as a basic form of easily communicable semiotics, even over a combination of gestures and vocalizations.[73] Gestures are embodied, enacted, and extended in that they create and distribute the cognitive process beyond the skin and out into space. Furthermore, the visuospatial aspect of the gestural process has been shown to activate the pre-motor sensory areas of the brain involved in the contemplation of movement and significant areas associated with the recognition of affective states in others. In this way perceived gestures enable another kind of kinesthetic empathy and also place concepts into the visuospatial realm. A large body of research now suggests how both "physically performed" and "simulated imagined" gestures influence human reasoning and problem solving.[74] Gestures have also been linked to the facilitation of abstract thought in that they help our minds move beyond basic cognitive functions as we "grasp" at thoughts and "reach out" for new ideas.

Clearly, gestures are an important part of the theatrical communication of both narrative and emotions, and provide audiences with visuospatial scaffolding with which to project and develop abstract thought. But our evidence for the gestures of Greek drama in practice is limited to their depictions on vase paintings and sculpture, some later rhetorical handbooks, and references in the texts.[75] While gestural sign systems are culturally specific and tend to operate in tandem with linguistic production, contextual information on ancient objects allows us to identify at least a partial typology of ancient gestures.[76] In addition, the texts of Greek drama contain many references to gestures, particularly in choral songs. A fine example can be seen in the choral entry song of Aeschylus's *Libation Bearers* (22–25).

> Sent from the House to bear libations
> Heavy hands, beating hard.
> Cheeks marked with crimson gashes,
> Nails plough furrows, fresh and deep.

We also have many portrayals of funerary rituals in ancient art, which depict this kind of beating of heads and chests and scratching of cheeks until they bleed, something impossible to show on a mask and here embodied in the lyrics of the song and probably enacted with precise and coordinated gestures of grief that were well known to the audience.[77]

One resource at our disposal already mentioned is a cross-cultural survey of living masked theatre traditions such as Kathakali, Noh, Topeng, and Commedia, all of which emphasize the importance of clarity in their respective gestural systems and the precision with which they must be performed.[78] In Noh, for example, there is an intricate movement system known as *kata* that is learned by the apprentice performer. For example, the *posture* called *shiori* denotes grief and involves the masked performer slightly lowering the head and drawing up one or both hands over the eyes. One hand can sometimes be used to point off into the distance to locate the cause of the grief in distal space.[79] *Kata* in Japanese Noh and Kabuki theatre can be likened to the *schemata* of Greek dance as described by Plutarch as the completion of a gesture or movement.[80] This idea of *schemata* has been viewed as some type of system of "fixed poses," but Llewellyn-Jones has shown how a fluidity of movement between gestures and dance steps probably operated in the ancient Greek theatre.[81] Plutarch also describes performers using movement to embody characters such as Apollo, Pan, or the Bacchae and connects poetry and movement by calling dancing "mute poetry" and poetry "speaking dancing." For Plutarch the sound of sung poetry "excites the hands and feet" and "raises every part of the entire body." We see something of this in Greek vase paintings where masked performers are frequently depicted in the act of movement with pronounced gestures with great attention to detail on the rendering of hands and feet (Figure 4.1).

There is also the text of Lucian's *On the Dance*, a dialogue where Lucian, a rhetorician from Roman Syria, attempts to convince his reluctant companion to attend a performance of a Pantomime dancer that was written in the mid-second century CE. Roman Pantomime, which flourished from the first to sixth century CE, usually took the form of one masked dancer, performing to sung lyrics on a tragic or mythological theme by a soloist or chorus, accompanied by musicians.[82] Intricate gestures and *cheironomy* (hand gestures) were an important part of the Pantomime dancer's repertoire. Lucian extols the beauty of Pantomime and tells how Nero gave as a present to a visiting king from Pontus a dancer, as his guest believed that his gestures were so effective in conveying non-verbal information that he could help him more effectively communicate with his neighbors who did not know his language (*On the Dance* 64). Lucian praises the expressivity of the Pantomime dancers' gestures and locates them as one of the most important aspects of the art form:

> Other arts call out only one half of a man's powers – the bodily or the mental: the pantomime combines the two. His performance is as much an intellectual as a physical exercise: there is meaning in his movements; every gesture has its significance; and therein lies his chief excellence.
>
> (Lucian, *On the Dance*, 69)

Like all body movements, gestures are amplified by the mask. In my own work in the theatre with masks, I have found that actors tend to instinctively compensate for their perceived lack of facial expression with pronounced gestures. A photograph from Desiree Sanchez's 2011 production of Pirandello's *Six Characters in Search of an Author* shows this in action (Chapter Three; Figures 3.4 and 3.5). If we compare these gestures to an Attic red-figure krater from Spina (Figure 4.1) that shows a boy holding a mask next to a masked actor playing a maenad, we see something similar. The gestures of the masked actor on the right are also highly pronounced with an emphasis placed in the hands and feet, which are the gestural visual focal points of the masked performer. The artist has also chosen to juxtapose the relaxed "off-stage" deportment of the actor on the left with the far more dynamic treatment of the masked actor on the right, with one foot raised, arm outstretched and one arm behind, foreshortened, denoting a turning movement. All of this is in complete contrast to the linear frontality of the "off-stage" actor's body. Just look at his feet.

Just as increasing literacy abilities in late-fifth-century Athens may well have affected how the mask was processed, we also see an aesthetic shift in theatrical gestures around this time. In the *Poetics*, Aristotle recounts a story of a distinguished tragic actor named Mynniskos, who had worked with Aeschylus, who insulted a younger actor, Kallippides, by calling him an ape. Aristotle tells us that this was because Kallippides "overdid everything" and in particular his gestures (1461b-62a). Csapo and Slater have proposed that compared to the "stately" gestures of Aeschylean drama, Kallippides' more naturalistic movements seemed "twitchy and hyperactive."[83] The idea of the more mannered style of Aeschylus in contrast to the new naturalism of Euripides is certainly found in Aristophanes. For example, it forms the basis of the competition between Aeschylus and Euripides in *Frogs* and in the costuming scene in *Acharnians* (404–434) when Dikaiopolis beseeches Euripides to dress him in the rags of one of his impoverished kings so he might elicit an empathetic response from the chorus. Just as we can detect an aesthetic shift toward a kind of naturalism in the works of Euripides, Aristophanes, and then Middle and New Comedy, we can also see a corresponding change in the form of the more stylized and schematic features of the mask and the gestures of the actor.[84]

Gestures have been shown to enhance the fluency of prediction generation by activating sensorimotor areas that help provide perceptual information.[85] Gestures also help off-load some of the cognitive effort involved in conceptualizations of stored memories and future possibilities. They are therefore more likely to be produced when there is a disconnect between the physical and mental environments, such as when a scene is being described that is not actually present, as in the theatre.[86] Speech is also more understandable in many circumstances when accompanied by gestures, which provide proprioceptive and kinematic information to both the speaker and the viewer. In the theatre, gestures not only assist the actor with memory recall but also add an important non-verbal level of sensorimotor communication to the performance.

Chorality and empathy

We have now examined the cognitive impact of performed stage movements and gestures and how the mask enhanced their communication and reception.

However, in the Greek theatre movement was not an individuated event; rather, much of it was performed within a broader cultural context of collective movement as represented by the chorus. Moreover, the chorus developed from one of the earliest forms of collective performance, the procession, and fundamental elements of processional performance are woven into the DNA of fifth-century drama. The chorus was a powerful mediating force in the communication of kinesthetic empathy, yet it remains one of the most misunderstood elements of ancient stage practice. We can help rectify this by viewing it within the cultural context of Greek processional culture and through the lens of group movement psychological studies.[87]

In modern parlance, the word "chorus" has come to denote a group of singers or the singer/dancers of a Broadway/West End musical, but in ancient Greek the term has several interrelated meanings that are all connected to the idea of group *movement*. Thus, it can mean dancers, the dancing place, and the thing that was danced.[88] In Athenian drama, the chorus both sung and danced, and the prominence of dance in ancient drama was reflected in the title of the wealthy citizen who received public acclaim for producing the play – the *choregos*, the leader of the dance. Greek theatrical choral dance was the manifestation in movement (often *with* song) of a highly organized presentational group interaction with performative roots in the procession and cult spectacle.

Choral singing and dancing was clearly a very ancient practice in the Greek world, and it was deeply woven into the fabric of Greek ritual and festive life. Even the mythical genesis of acting is imagined as a development of an existing choral art form. For example, in the sixth-century *Hymn to Apollo*, we can read of a chorus of Delian girls singing about taking on the vocal personas of the characters in their song.[89] Likewise, Thespis, the legendary first actor, was said to have stepped out of a chorus to answer them. No classical Greek play was without a chorus, and in the Athenian theatre the chorus was an almost-continuous presence in the spectator's visual field with 12–15 members in tragedy and satyr play and 24 in comedy. We should also remember that the opening day of the Festival of Dionysos was the dithyrambic competition with choruses 50 strong recruited from the tribes of Attica. In drama, the chorus also created another level of contrast between peripheral and foveal vision as they moved from the centrality of their odes to the periphery of the actor scenes and back again. Yet, the chorus was not "off stage" or even diminished when they were not fully engaged in their dances and songs, which is how we encounter them in the texts. Instead, their perpetual masked physical presence, silently listening and observing, and even occasionally audibly reacting to the events taking place before them, kept the emotional force and narrative direction of the play in both areas of the spectator's visual field. Even during an actor's speech or a dialogue scene, the gaze direction of the chorus and their physical actions would have done a great deal to contribute to the emotional intensity of what was being presented on stage, just as the group movements, dance steps, and gestures that accompanied their songs would have magnified the interactive communication of kinesthetic empathy. The attitude of the Greek audience to the choruses they watched was informed by their intensive participation within a wider cultural context of chorality and procession, and the latter has been relatively unexplored as a major inspiration on the stage dynamics and reception of ancient drama.

Across many diverse cultures, processions are one of the earliest forms of institutional public performance. Victor Turner placed the procession as one of the oldest forms of group performance that are a public expression of *communitas* – a state of performative equality that fostered shared experience and was often combined with a ritual of liminality and transformation.[90] The performance theorist Richard Schechner divides such early performance forms into two broad categories: "eruptions" and "processions."[91] An "eruption" is a static event that unfolds in one location where a crowd gathers to watch. Conversely, a procession has a predetermined route and a fixed, final goal, and it follows an organized structure and a commonly understood form. Hence, the visual displays inherent in the procession are very important in communicating identity, status, and power. Schechner describes how the procession has a tendency to make several stops along its route where associated stationary performances take place. These are processional "eruptions" and spectators can gather to watch, participate, and/ or continue to follow the procession to its ultimate goal."[92] Thus, participation at a procession is far more active than at an "eruption," as processions by their very nature usually require a large number of participant/performers in order to achieve their aim of providing a spectacle suitable of transforming the territory they cross from the everyday to the extraordinary. Spectatorship at a procession is also mediated by constantly shifting boundaries between performance forms. For example, the spectators may watch a procession by placing themselves in a static viewing position and observe it passing. Alternatively, they might shift their position and move to one of the several stops on the procession route in order to watch a standing performance (in ancient processions this would usually be a song, dance, or a dance-play). Furthermore, as the processional form encourages a general movement by the spectators to observe the several static performances staged along the route, spectators could find themselves swept up by the motion of the procession, moving with it from point to point and on to its final destination.

One can transpose these concepts onto the fifth-century Athenian theatre, which acted as a terminus for the Dionysian procession for sacrificial offerings and performances in honor of the god. The form of the theatre retains a physical memory of its processional origins in that its two opposing side entrances are called *eisodoi* – "side roads" – and the spectators are seated in the *theatron* – a viewing place that in the fifth century resembled the temporary grandstands that were erected in the Agora and elsewhere for processions, athletic competitions, and other performance events. The theatrical choruses that used these processional routes to make their entrances were part of this continuity between active participation in a procession and experiencing drama as a spectator.

That the Athenians regarded drama as closely linked to the procession can be seen in the "law of Euagoras," quoted by Demosthenes in the fourth century, granting legal amnesty during certain religious festivals:

> Euagoras moved: whenever there is the procession for Dionysos in Piraeus and comedy and tragedy, whenever there is a procession at the Lenaion and tragedy and comedy, whenever there is at the City Dionysia the procession and the boys <dithyramb> and the *komos* and comedy and tragedy, and whenever there is a procession at the Thargelia. It shall not be permitted to take security or to arrest another, not even those past-due their payments during these days.
>
> (Demosthenes, *Against Meidias* 10, tr. adapted from Csapo and Slater)

It is notable that these three festivals to Dionysos and one to Apollo (the Thargelia) are described in terms of the *pompe* (procession), and although tragedy and comedy are referenced, the procession stands out as the central descriptive element for these performing-arts festivals. Walter Burkert describes the procession (*pompe*) as "the fundamental medium of group formation" and writes that "hardly a festival is without a *pompe*."[93] Processions accompanied sacrifices, and they were the most public aspect of festivals; they transported worshippers to sacred shrines, accompanied the idols of gods, and conveyed initiation rites. That processions were the preeminent performance form in Athens is borne out by a survey of Robert Parker's extensive appendix of Athenian festivals. He produces a list of 39 annual processional events that we know of that took place in the city.[94]

The main dramatic festival in Athens, the City Dionysia, opened with a large procession, open to foreign delegations from the allied states who processed with a phallus, in honor of the god.[95] This Dionysian procession was second in scale only to the great Panathanaea, and its culmination point was the Sanctuary of Dionysos Eleuthereus on the southeast slope of the Acropolis, where at least 100 animals were ritually slaughtered before the temple just beneath the location of the *theatron*.[96] The exact program of performances fluctuated throughout the fifth and fourth century, but the basic model included at least a day for twenty 50-person dithyrambic choral performances, and then there were three or four days of tragedies, each culminating in the performance of a satyr play, and then a full day for the presentation of five comedies, although later when this was reduced to three the comedies may also have been presented individually on the tragedy days after the satyr performance.[97] At the end of the festival, the victorious parties were celebrated with their own ribald processions that led participants from the theatre to private houses.

Processional choral movement was the main kinetic form of the entire festival. For example, the dithyrambs were delivered by a large 50-strong choral group that processed into the theatre and danced in unison as it sang; both tragedy and comedy were also predominantly choral performance forms and developed to inhabit a space specifically intended for the presentation of processional drama.

Furthermore, the narratives of both tragedy and comedy include the staging of numerous processional events interwoven within the plots of the plays, such as choral entrances, wedding marches, funeral processions, or sacrificial rituals. One notable example is the ending of Aeschylus's *Oresteia*, where the chorus of Furies are invested as "Kindly Ones" by means of a procession reflective of both the Dionysia and Panathanaea processions. In comedy, processions are re-created for comic effect, such as Dicaeopolis's "Dionysia" in *Acharnians* (241–262), the mini "Panathenaea" in *Ecclesiazousae* (730–756), and the "wedding" procession of Peithetairos and Basiliea in *Birds* (1706–1765). Comedies often used processional and dance forms to create movement-filled finales.

Collective movement in a presentational and processional group forms spatial configurations that define relationships between individual people, individuals and groups, and different groups. In effect, they are one of the fundamental ways in which performance creates the kind of interactive bodily enacted "we-space" where mutual co-presence and engagement is paramount. Movement analyst Irmgard Bartenieff noted that these movement configurations outline the territory where action-interaction develops and communicates what that action-interaction might become. The individuals within a certain choral group might be placed in files, rows, circles, or a variant of them, and this basic configuration "will be critical to the nature of their confrontations with each other and of the confrontations of their group with another group."[98] For example, Bartenieff regarded the file, where participants line up directly behind each other, as a predominantly passive movement form for an individual within a group with minimal interaction. It is always deliberately chosen rather than organic, and it is used with prisoners or slaves as a configuration of compliance and control. On the other hand, the row, where participants are placed side by side, provides "an interrelationship of equality" and encourages the sharing of the same action and group focus. The row, the form that has been most often associated with the tragic chorus, projects solidarity and enables both advancing and retreating, "mutual reinforcement," and an unbroken line against an intruder.[99]

Bartenieff also noted that a row can most easily lead to a circle. The circle, which is often associated with the dithyramb, promotes group cooperation and collective sharing, as there is a universal relationship to the center when the group faces it and therefore each other. D'Angour has also identified the strengths of the circular dance form for creating group unity, especially in the case of dance with vocal delivery.[100] He suggests that the dithyramb may have originally been a processional event that was organized into a circular form by Lassos of Hermione around the end of the sixth century. This is perhaps why Pindar's *Dithyramb 2* starts by referring to how the dithyramb was once "stretched out like a rope" but now the chorus dances "in a circle." D'Angour's theory does highlight what seems to be a marked development from processional to stationary performance. This was probably due to a reorganization of the festival and the establishment of the Sanctuary of Dionysos Eleuthereus around 530. Furthermore, the circle-dance was not only associated with the dithyramb but was also a feature of the dance performance of lyric as opposed to the rectangular form of tragic dancing.[101]

In addition to the side-to-side contact offered by the row, "the shared relation-ship with a center makes body, space and inter-actions more synchronous."[102] The circle, where the individual members of the group face inward, promotes the development of a common rhythm as steps are transmitted equally from side-to-side and across the center of the circle, even to the individual who may be farthest away, thus unifying the group. "In the circle, Effort Flow most easily helps estab-lish the common continuity of the movement and the common order of step direc-tion. A circle thus brings people together, it is one of the oldest forms of social congregation in dance."[103] It is notable then that where processions halt at key locations for static performances, the circle dance tends to prevail.[104] Bartenieff's work also examined the group movement dynamics of spontaneous improvisa-tions, and she concluded that they often contained elements of confrontations that appear in communities. Their performance by the group helps reinforce accept-able rules of community interrelationships. She noted that when they do break out and momentarily threaten the harmony of the group, the tension is often mitigated by the emergence of a common "effort rhythm" that spreads throughout the entire group and helps to reinforce its solidarity. This is kinesthetic empathy in practical movement-based terms and is directly applicable to the collective movements of the Greek dramatic chorus and how they were received by an audience immersed in a processional performance culture.

In *Laws*, Plato finds no real division between dances and processions and posits that the ability to create ordered movement in a chorus is given by the gods and distinguishes men from animals (653e).[105] He goes on to make the famous state-ment that the uneducated man is "without the dance" (654a). For the spectator watching the procession or dance, Judith Lynne Hanna has stated, "motion has the strongest visual appeal to attention for it implies a change in the conditions of the environment which may require action. Used extraordinarily in the dance, motion is potently related to the experience of arousal and motivation."[106] This is directly applicable to collective movement forms such as the procession and cho-ral drama that seek to transform their respective environments via the use of group movement, visuality, dance, music, and rhythm. Hanna goes on to point out that in dance, the motor/visual-kinesthetic channels predominate instead of the vocal/auditory channels, in that language exists in a temporal dimension, whereas dance involves the temporal plus the three dimensions of space. Thus, the relationship of a procession to the space it moves through is an essential feature that links the visual display to its environment, both ritualizing the city streets and visiting loca-tions of religious and civic significance to imbue the event with additional power.

The processional performance form was not only one of public display and group participation, but it was also a fundamental element in Greek telestic (*initiation*) rites such as the Orphic rituals, the Corybantic "frenzies," and the Eleusinian mysteries, where participating in a procession, often in some sort of costuming, set to otherworldly and strange music (more on that in the next chap-ter), with rhythmic and repetitive group movement could produce a feeling of *catharsis* (purgation) in the initiate. One of the functions of these processions was to re-create the identity of the group within the senses of *enthusiasmos* and ecstasy. Thus, Andrew Ford has described the processions of Greek mystery cults as dramatic and highly theatrical, citing Aristides Quintilianus:[107]

> Accordingly, they say that there is a certain logic to Bacchic and similar rites whereby the feelings of anxiety . . . are cleared away through the melodies and dances of the ritual in a joyful and playful way.[108]

Masking also occurred in many of these ritual practices, including the monster masks worn by the initiates of Eleusis and used to terrify the procession on its way from Athens, the smearing of clay on faces, and the wearing of wreaths in Bacchic and other rites.[109] The mask, which probably originated in ritual performance, also acted as a material anchor, simultaneously acting as a referent of cultic action and helping to create a dramatic character. It enabled the chorus to present itself both as a unified group but also as individuals within that group perhaps with different concerns from each other. This would have provided the ancient dramatist with a visual panoply of differing opinions expressed by the movement of a gesture, the tilt of a masked head, or the subtle aspect of the fingers showing tension, pleasure, or pain.[110] Textually we can detect differences of opinions within dramatic choral groups, such as the 12 distinct voices we see reacting to the death of Agamemnon in *Agamemnon* (1343–1372), but the visual display afforded by masks and movement must have offered an even greater range of responses and emotional shading.

One final, but important point about the movement of the dramatic chorus. Many, including the dance scholar Lawler, have tended to portray Greek drama, especially tragedy, as "stately and dignified" in its movements.[111] But the influences on and the evidence within the texts of the plays does not support this old-fashioned view that sadly still tends to prevail in the popular imagination of Greek drama. Graham Ley has rightly summed up the situation:

> To conclude against movement and dancing would be to insist on the verbal apart from the physical without good reason and, perhaps, to contradict the express purpose of composition for tragic performance by *choros* and actors.[112]

We should also lay to rest the idea that dramatic choral odes are "interludes" that interrupted the actors' scenes to comment back upon them. Greek dramas were staged with actor scenes woven between choral songs in the overall narrative structure of the plays, and it was rare for the chorus to leave the playing space once they had entered. Their presence would be a constant reactive emotive agent. Thus, the chorus were part of a seamless expression of a bi-modal peripheral and foveal spectator gaze; they added nuance, complexity, and range to the emotional responses presented, and they radiated and magnified the kinesthetic, empathetic communication attributes of the performance. Instead of thinking about Greek plays as scenes with choral interludes, we should instead view them as choral performances dramatically interrupted by scenes.[113]

The movement of Greek drama was a vital element in the total performance experience and bodily affected those who saw it. Its *dianoia* was the dynamic intentionality it brought to the enactment narrative and the non-verbal communication power between masked performer and spectator. Most of this movement was set to music, and yet we know almost nothing about what the music of Greek drama actually sounded like. In the next chapter we will take a cognitive approach to the little we do know and try to gain a better understanding of how the distinctive music and instrumentation of the Greek theatre affected, dissociated, and absorbed those who heard it performed.

Notes

1 Nietzsche 1993: 46.
2 Nietzsche 1993: 23.
3 Henrichs 1994: 56–111.
4 Clark 2015: 66.
5 Clark 2015: 7.
6 On *skenographia* (scene painting), see Davidson 2005; Csapo and Slater 1995: 258 and 273–274; Padel 1990: 346–354.
7 On the iconography of Greek costume, see Wyles 2011.
8 On the staging of this scene, see Heath and OKell 2007: 363–370 and Finglass 2011: 135–174 and Scullion 1994.
9 Aristophanes, *Frogs* 304. On Hegelochus's mistake, see Farmer 2016: 31–34, 86–87.
10 Janko 1987: 117 and n, on 55a29.
11 Aristophanes, *Archanians* 412 and *Thesmophoriazusae* 148–152. See also Plato, *Ion* 535B-C.
12 Also found in Sophocles' Electra 920, 1199, and 1200; and Euripides' Electra 545 and Hecuba 341.
13 Webb 2008: 60.
14 For example, the *Watching Dance: Kinesthetic Empathy* research project. www.watch ingdance.org/. Accessed August 31st 2016.
15 Hatfield et al. 1994.
16 Verbeke 1997.
17 Houser and Waldbuesser 2016: 1–8.
18 Rueff-Lopes et al. 2015: 412–434.
19 Kramer et al. 2014: 8788–8790.
20 Kleinsman and Buckley 2015: 179–182.
21 Ondobaka et al. 2015: 172–188.
22 Ondobaka et al. 2015: 180.
23 Fogtmann 2007.
24 Calvo-Merino et al. 2008: 911–922.
25 Shamay-Tsoory et al. 2009: 617–627, Hogeveen et al. 2016: 694–705.
26 Harris 2007: 203–231.
27 di Pellegrino et al. 1992: 176–180; Rizzolatti and Craighero 2005.
28 Iacoboni 2008.
29 Cook et al. 2014: 177–192; Caramazza et al. 2014.
30 Corradini and Antonietti 2013: 1152–1161.
31 McGarry and Russo 2011: 178–184.
32 Avenanti et al. 2005: 955–960.
33 Wicker et al. 2003: 655–664.
34 Trevarthen and Fresquez 2015: 194–210.

35 Clark 2015: 151–152.
36 Clark 2015: 151.
37 Kilner and Friston 2014: 207–208.
38 On Theory of Mind applied to ancient comedy, see Ruffell 2008: 45–50.
39 Johansson 1973: 201–211.
40 Friston et al. 2010: 227–260.
41 Clark 2015: 123.
42 Gallagher 2005: 65–85.
43 Ruth Webb explores the masked Roman pantomime dancer from the perspective of the performer and describes the "unthinking" quality of Kathakali movement training to this ancient art form. Webb 2008: 43–60.
44 Krueger and Michael 2012.
45 Jola 2010: 203–204.
46 Calvo-Merino et al. 2005:1243–1249. See also Calvo-Merino et al. 2006: 1905–1910; Brown et al. 2006: 1157–1167.
47 Aristophanes, *Frogs,* 729; Plato, *Laws* 7.814e – 817e. See also, Ley 2010: 150–166.
48 Skoyles 2008.
49 Zaki and Ochsner 2013: 214.
50 Niedenthal et al. 2005: 184–211.
51 For example, Zieber et al. 2014: 675–684; de Gelder and Hortensius 2014: 153–164; Hess and Fischer 2014: 45–57; Meeren et al. 2013; Christensen and Calvo-Merino 2013: 76.
52 Sugita 2009: 39–44.
53 Bogart et al. 2010: 134–142.
54 Van Rysewyk 2011.
55 Baylor 2009: 3559–3565.
56 de Gelder 2009: 3475–3484.
57 de Gelder 2010: 513–527.
58 Peelen and Downing 2005: 603–608.
59 Van den Stock et al. 2007: 487.
60 de Gelder 2009: 3475–3476.
61 Emigh 1996: 110; see also Emigh 2003.
62 Brecht 2014: 198.
63 Paraskeva 2013: 10.
64 Grotowski 2012: 142.
65 Schechner 2004: 201.
66 Bharata and Rangacharya 2003: 41.
67 Warden 2012: 66.
68 Copeland 2004: 112–113.
69 Matsumoto and Hwang 2013: 1–27.
70 Hutchins 2006: 376.
71 Malafouris 2013: 60.
72 Lakoff and Johnson 1999: 77.
73 Fay et al. 2013.
74 Alibali et al. 2014: 150.
75 On the problems with the evidence for gestures in Greek drama, particulary conflating gestural systems from other genres with dramatic gestures, see Valakas 2002 and Monaghan 2007.
76 Krueger and Michael 2012.
77 For an excellent survey of Roman gestural systems, which also includes a good deal of material on Greek gestures, see Corbeill 2004.
78 Although not masked, the dancers in European *ballet d'action* re-created the stories of classical mythology with gestural movement and dance. Lada-Richards connects this genre with Roman Pantomime dancing. Lada-Richards 2013: 163–174.

79 See Nakanishi and Kiyonari 1983; and Perzynski 2012: 176.
80 Plutarch *Quest. Conv.* 9.15.
81 Lawler 1954.
82 On Roman Pantomime, see Hall and Wyles 2008; Jory 2008: 157–168, Macintosh 2010; Lada-Richards 2013.
83 Csapo and Slater 1995: 256–257.
84 Aristotle credits Euripides with the first use of a more "naturalistic" style in drama (Rhetoric 1404b).
85 Glenberg and Gallese 2012: 905–922.
86 Pouw and Hostetter 2016: 57–80.
87 Recent attempts to rectify this include Billings et al. 2013 and Gagné and Hopman 2013.
88 The term *orchestra* for "dancing place" is first found in Aristotle *Prior Analytics* 901b30.
89 *Homeric Hymn to Apollo* 156–164.
90 Turner 1979: 465–499.
91 Schechner 2004: 153–186.
92 Schechner 2004: 159–160.
93 Burkert 1985: 99.
94 Parker 2005: 456–487.
95 Inscription from the colony of Brea *IG* 1ˢ 46.11–13. 446/5
96 Rehm 2002: 45; Cole 1993: 25–38.
97 For the various possible performance schedules, see Csapo and Slater 1995: 106–108.
98 Bartenieff and Lewis 1980: 130.
99 On the configurations of ancient choral groups, see Calame 2001: 34–35.
100 D'Angour 1997: 331–351.
101 See Calame 2001: 34–35. Circle dances are found depicted on Achilles' shield in the *Iliad* (18.504), performed by Phaeacian boys in the *Odyssey* (8.250), led by Theseus around the altar at Delos in *The Hymn to Delos* of Callimachus (300–316), and frequently associated with the Delian Maidens (Euripides' *Iphigenia in Taurus* 427–430, *Iphigenia at Aulis* 1054–1057, and *Herakles* 687–690). Heyschius describes the chorus in terms of a circle or crown (s.v. chorus X 645 Schmidt), and Callimachus imagines the islands of the Cyclades surrounding Delos like a chorus (*Hymn to Delos* 300–301).
102 Bartenieff and Lewis 1980: 132.
103 Bartenieff and Lewis 1980: 132.
104 This can still be observed as the Skyrian *Apokries*, a spring festival involving costumes, procession, dances, and song. See Caracusi and Bonanzinga 2008: 173–190.
105 See Lonsdale 2000: 41.
106 Hanna 1987: 75.
107 Ford 2016: 23–41.
108 Aristides Quintilianus 3.25.14–19.
109 Demosthenes *On the Crown* 18.259–260.
110 On the relationship between gestures, communication, and cognition, see McNeill 1992.
111 Lawler 1964.
112 Ley 2010: 91.
113 Aristotle made a distinction between the chorus and the rest of the play, coining the term *epeisodion* (episode) to describe the part of tragedy between choral songs (*Poetics* 1452b20–21). See Halleran 2005 and Taplin 1977: 470–476. In a workshop with theatre lecturers at the Association of Theatre in Higher Education conference in Los Angeles in July 2010, I placed 12 people in a half-circle around a masked performer and asked them to just watch, listen, and gently respond without "pulling focus." The result was mesmerizing for both spectators and participants and all agreed that the text spoken was enhanced by having this added visual dimension.

Bibliography

Alibali, M.W., Boncoddo, R. and Hostetter, A.B. 2014. "An embodied perspective" in Shapiro, L. (Ed.), *The Routledge handbook of embodied cognition.* London: Routledge: 150–159.

Avenanti, A., Bueti, D., Galati, G. and Aglioti, S.M. 2005. "Transcranial magnetic stimulation highlights the sensorimotor side of empathy for pain" in *Nature Neuroscience* 8: 955–960.

Bartenieff, I. and Lewis, D. 1980. *Body movement: Coping with the environment.* Hove, UK: Psychology Press.

Baylor, A.L. 2009. "Promoting motivation with virtual agents and avatars: Role of visual presence and appearance" in *Philosophical Transactions of the Royal Society of London B: Biological Sciences* 364.1535: 3559–3565.

Bharata, M. and Rangacharya, A. 2003. *The Natyasastra.* New Delhi, India: Munshiram Manoharlal Publishers.

Billings, J., Budelmann, F. and Macintosh, F. (Eds.). 2013. *Choruses, ancient and modern.* Oxford: Oxford University Press.

Bogart, K.R. and Matsumoto, D. 2010. "Living with Moebius syndrome: Adjustment, social competence, and satisfaction with life" in *The Cleft Palate-Craniofacial Journal* 47.2: 134–142.

Brecht, B. 2014. *Brecht on theatre.* London: Bloomsbury.

Brown, S., Martinez, M.J. and Parsons, L.M. 2006. "The neural basis of human dance" in *Cerebral Cortex* 16.8: 1157–1167.

Burkert, W. 1985. *Greek religion.* Cambridge, MA: Harvard University Press.

Calame, C., Collins, D. and Orion, J. 2001. *Choruses of young women in ancient Greece: Their morphology, religious role, and social functions.* Lanham, MD: Rowman & Littlefield.

Calvo-Merino, B., Glaser, D.E., Grezes, J., Passingham, R.E. and Haggard, P. 2005. "Action observation and acquired motor skills: An FMRI study with expert dancers" in *Cerebral Cortex* 15.8: 1243–1249.

Calvo-Merino, B., Grèzes, J., Glaser, D.E., Passingham, R.E. and Haggard, P. 2006. "Seeing or doing? Influence of visual and motor familiarity in action observation" in *Current Biology* 16.19: 1905–1910.

Calvo-Merino, B., Jola, C., Glaser, D.E. and Haggard, P. 2008. "Towards a sensorimotor aesthetics of performing art" in *Consciousness and Cognition* 17.3: 911–922.

Caracusi, M.R. and Bonanzinga, S. 2008. "Carnevale a Skyros" in *Archivo Anthropologico Mediterraneo* 10: 173–190.

Caramazza, A., Anzellotti, S., Strnad, L. and Lingnau, A. 2014. "Embodied cognition and mirror neurons: A critical assessment" in *Annual Review of Neuroscience* 37: 1–15.

Christensen, J.F. and Calvo-Merino, B. 2013. "Dance as a subject for empirical aesthetics" in *Psychology of Aesthetics Creativity and the Arts* 7.1: 76–88.

Clark, A. 2015. *Surfing uncertainty: Prediction, action, and the embodied mind.* Oxford: Oxford University Press.

Cole, S.G. 1993. "Procession and celebration at the Dionysia" in Scodel, R. (Ed.), *Theater and society in the classical world.* Ann Arbor, MI: University of Michigan Press: 25–38.

Cook, R., Bird, G., Catmur, C., Press, C. and Heyes, C. 2014. "Mirror neurons: From origin to function" in *Behavioral and Brain Sciences* 37.2: 177–192.

Copeland, R. 2004. *Merce Cunningham: The modernizing of modern dance.* London: Routledge.

Corbeill, A. 2004. *Nature embodied: Gesture in ancient Rome.* Princeton, NJ: Princeton University Press.

Corradini, A. and Antonietti, A. 2013. "Mirror neurons and their function in cognitively understood empathy" in *Consciousness and Cognition* 22.3: 1152–1161.

Csapo, E. and Slater, W.J. 1995. *The context of ancient drama*. Ann Arbor, MI: University of Michigan Press.

D'Angour, A. 1997. "How the dithyramb got its shape" in *The Classical Quarterly (New Series)* 47.2: 331–351.

Davidson, J. 2005. "Greek drama: Image and audience, the TBL webster memorial lecture 2003" in *Bulletin of the Institute of Classical Studies* 48.1: 1–13.

Emigh, J. 1996. *Masked performance: The play of self and other in ritual and theatre*. Philadelphia, PA: University of Pennsylvania Press.

Emigh, J. 2003. "Playing with the past: Visitation and illusion in the mask theatre of Bali" in Schechter, J. (Ed.), *Popular theatre: A sourcebook*. London/New York: Routledge: 107–128.

Farmer, M.C. 2016. *Tragedy on the comic stage*. Oxford: Oxford University Press.

Fay, N., Lister, C.J., Ellison, T.M. and Goldin-Meadow, S. 2013. "Creating a communication system from scratch: Gesture beats vocalization hands down" in *Frontiers in Psychology* 5: 354–354.

Finglass, P.J. (Ed.). 2011. *Sophocles: Ajax* (Vol. 48). Cambridge: Cambridge University Press.

Fogtmann, M.H. 2007. "Kinesthetic empathy interaction – exploring the possibilities of psychomotor abilities in interaction design" in Ramduny-Ellis, D., Dix, A. and Gill, S. (Eds.), *Second International Workshop on Physicality in proceedings of the 21st British HCI Group Annual Conference on People and Computers: HCI . . . but not as we know it*. London: British Computer Society: 217–218.

Ford, A.L. 2016. "Catharsis, music, and the mysteries in Aristotle" in *Skenè: Journal of Theatre and Drama Studies* 2.1: 23–31.

Friston, K.J., Daunizeau, J., Kilner, J. and Kiebel, S.J. 2010. "Action and behavior: A free-energy formulation" in *Biological Cybernetics* 102.3: 227–260.

Gagné, R. and Hopman, M.G. 2013. *Choral mediations in Greek tragedy*. Cambridge: Cambridge University Press.

Gallagher, S. 2005. *How the body shapes the mind*. Oxford: Clarendon Press.

Gelder, B. de. 2009. "Why bodies? Twelve reasons for including bodily expressions in affective neuroscience" in *Philosophical Transactions of the Royal Society B: Biological Sciences* 364.1535: 3475–3484.

Gelder, B. de and Hortensius, R. 2014. "The many faces of the emotional body" in Christen, Y. and Decety, J. (Eds.), *New frontiers in social neuroscience*. New York: Springer International Publishing: 153–164.

Gelder, B. de, Van den Stock, J., Meeren, H.K., Sinke, C.B., Kret, M.E. and Tamietto, M. 2010. "Standing up for the body: Recent progress in uncovering the networks involved in the perception of bodies and bodily expressions" in *Neuroscience & Biobehavioral Reviews* 34.4: 513–527.

Glenberg, A.M. and Gallese, V. 2012. "Action-based language: A theory of language acquisition, comprehension, and production" in *Cortex* 48.7: 905–922.

Grotowski, J. 2012. *Towards a poor theatre*. London: Routledge.

Hall, E. and Wyles, R. (Eds.). 2008. *New directions in ancient pantomime*. Oxford: Oxford University Press.

Halleran, M.R. 2005. "Episodes", in Gregory, J. (Ed.), *A companion to Greek tragedy*. Malden, MA: Blackwell: 167–182.

Hanna, J.L. 1987. *To dance is human: A theory of nonverbal communication*. Chicago: University of Chicago Press.

Harris, D.A. 2007. "Pathways to embodied empathy and reconciliation after atrocity: Former boy soldiers in a dance/movement therapy group in Sierra Leone" in *Intervention* 5.3: 203–231.

Hatfield, E., Cacioppo, J.T. and Rapson, R.L. 1994. *Emotional contagion*. Cambridge: Cambridge University Press.

Heath, M. and OKell, E. 2007. "Sophocles' Ajax: Expect the unexpected" in *The Classical Quarterly (New Series)* 57.2: 363–380.

Henrichs, A. 1994. "'Why should I dance?' Choral self-referentiality in Greek tragedy" in *Arion: A Journal of Humanities and the Classics* 3.1: 56–111.

Hess, U. and Fischer, A. 2014. "Emotional mimicry: Why and when we mimic emotions" in *Social and Personality Psychology Compass* 8.2: 45–57.

Hogeveen, J., Salvi, C. and Grafman, J. 2016. "'Emotional intelligence': Lessons from Lesions" in *Trends in Neurosciences* 39.10: 694–705.

Houser, M.L. and Waldbuesser, C. 2016. "Emotional contagion in the classroom: The impact of teacher satisfaction and confirmation on perceptions of student nonverbal classroom behavior" in *College Teaching*: 1–8.

Hutchins, E. 2006. "The distributed cognition perspective on human interaction" in Schegloff, E.A. (Ed.), *Roots of human sociality: Culture, cognition, and interaction* 1. Oxford: Berg Publishers: 375–398.

Iacoboni, M. 2008. *Mirroring people*. New York: Farrar, Straus & Giroux.

Janko, R. 1987. *Aristotle: Poetics*. Cambridge, MA/Indianapolis: Hackett Publishing.

Johansson, G. 1973. "Visual perception of biological motion and a model for its analysis" in *Perception & Psychophysics* 14.2: 201–211.

Jola, C. 2010. "Research and choreography: Merging dance and cognitive neuroscience" in Blaesing, B., Puttke, M. and Schack, T. (Eds.), *The neurocognition of dance: Mind, movement and motor skills*. London: Psychology Press: 203–234.

Jory, J. 2008. "The pantomime dancer and his libretto" in Hall, E. and Wyles, R. (Eds.), *New directions in ancient pantomime*. Oxford: Oxford University Press: 157–168.

Kilner, J.M. and Friston, K.J. 2014. "Relating the 'mirrorness' of mirror neurons to their origins" in *Behavioral and Brain Sciences* 37.2: 207–208.

Kleinsman, J. and Buckley, S. 2015. "Facebook study: A little bit unethical but worth it?" in *Journal of Bioethical Inquiry* 12.2: 179–182.

Kramer, A.D., Guillory, J.E. and Hancock, J.T. 2014. "Experimental evidence of massive-scale emotional contagion through social networks" in *Proceedings of the National Academy of Sciences* 111.24: 8788–8790.

Krueger, J. and Michael, J. 2012. "Gestural coupling and social cognition: Möbius Syndrome as a case study" in Frontiers in Human Neuroscience 6.81. doi: 10.3389/fnhum.2012.00081

Lada-Richards, I. 2013. *Silent eloquence: Lucian and pantomime dancing*. London: Bloomsbury.

Lakoff, G. and Johnson, M. 1999. *Philosophy in the flesh: The embodied mind and its challenge to western thought*. New York: Basic books.

Lawler, L.B. 1954. "Phora, schêma, deixis in the Greek dance" in *Transactions and proceedings of the American philological association* (Vol. 85). Johns Hopkins University Press, MD: American Philological Association: 148–158.

Lawler, L.B. 1964. *The dance of the ancient Greek theatre*. Iowa City, IA: Univesity of Iowa Press.

Ley, G. 2010. *The theatricality of Greek tragedy: Playing space and chorus*. Chicago: University of Chicago Press.

Lonsdale, S.H. 1993. *Dance and ritual play in Greek religion*. Baltimore, MD: Johns Hopkins University Press.

Lonsdale, S.H. 2000. *Dance and ritual play in Greek religion*. Baltimore, MD: Johns Hopkins University Press.

McGarry, L.M. and Russo, F.A. 2011. "Mirroring in dance/movement therapy: Potential mechanisms behind empathy enhancement" in *Arts in Psychotherapy* 38: 178–184.

Macintosh, F. 2010. *The ancient dancer in the modern world: Responses to Greek and Roman dance*. Oxford: Oxford University Press.

McNeill, D. 1992. *Hand and mind: What gestures reveal about thought*. Chicago: University of Chicago press.

Malafouris, L. 2013. *How things shape the mind*. Cambridge, MA: MIT Press.

Matsumoto, D. and Hwang, H.C. 2013. "Cultural similarities and differences in emblematic gestures" in *Journal of Nonverbal Behavior* 37.1: 1–27.

Meeren, H.K., Gelder, B. de, Ahlfors, S.P., Hämäläinen, M.S. and Hadjikhani, N. 2013. "Different cortical dynamics in face and body perception: An MEG study" in *PLoS One* 8.9: e71408.

Monaghan, P. 2007. "Mask, word, body and metaphysics in the performance of Greek tragedy" in *Didaskalia* 7.1.

Nakanishi, T. and Kiyonari, K. 1983. *Noh Masks*. Kenny, D. (Tr.). Osaka, Japan: Hoikusha Publishing Company.

Niedenthal, P.M., Barsalou, L.W., Winkielman, P., Krauth-Gruber, S. and Ric, F. 2005. "Embodiment in attitudes, social perception, and emotion" in *Personality and Social Psychology Review* 9.3: 184–211.

Nietzsche, F. 1993. *The birth of tragedy: Out of the spirit of music*. London: Penguin.

Ondobaka, S., Kilner, J. and Friston, K. 2015. "The role of interoceptive inference in theory of mind" in *Brain and Cognition*: 112: 64–68.

Padel, R. 1990. "Making space speak" in Zeitlin, F.I. and Winkler, J.J. (Eds.), *Nothing to do with Dionysos? Athenian drama in its social context*. Princeton, NJ: Princeton University Press: 336–365.

Paraskeva, A. 2013. *The speech-gesture complex: Modernism, theatre, cinema*. Edinburgh: Edinburgh University Press.

Parker, R. 2005. *Polytheism and society at Athens*. Oxford: Oxford University Press.

Peelen, M.V. and Downing, P.E. 2005. "Selectivity for the human body in the fusiform gyrus" in *Journal of Neurophysiology* 93.1: 603–608.

Pellegrino, G. di, Fadiga, L., Fogassi, L., Gallese, V. and Rizzolatti, G. 1992. "Understanding motor events: A neurophysiological study" in *Experimental Brain Research* 91.1: 176–180.

Perzynski, F. 2012. *Japanese no masks: With 300 illustrations of authentic historical examples*. Mineola, NY: Courier Corporation.

Pouw, W. and Hostetter, A.B. 2016. "Gesture as predictive action" in *Reti, Saperi, Linguaggi* 3.1: 57–80.

Rehm, R. 2002. *The play of space: Spatial transformation in Greek tragedy*. Princeton, NJ: Princeton University Press.

Rizzolatti, G. and Craighero, L. 2005. "Mirror neuron: A neurological approach to empathy" in Changeux, J.P.P., Damasio, A. and Singer, W. (Eds.), *Neurobiology of human values*. Berlin/Heidelberg/New York: Springer: 107–123.

Rueff-Lopes, R., Navarro, J., Caetano, A. and Silva, A.J. 2015. "A Markov chain analysis of emotional exchange in voice-to-voice communication: Testing for the mimicry hypothesis of emotional contagion" in *Human Communication Research* 41.3: 412–434.

Ruffell, I. 2008. "Audience and emotion in the reception of Greek drama" in Revermann, M. and Wilson, P. (Eds.), *Performance, iconography, reception: Studies in honour of Oliver Taplin*. Oxford: Oxford University Press: 37–58.

Schechner, R. 2004. *Performance theory*. London: Routledge.

Scullion, S. 1994. *Three studies in Athenian dramaturgy* (Vol. 25). Stuttgart/Leipzig, Germany: Walter de Gruyter.

Shamay-Tsoory, S.G., Aharon-Peretz, J. and Perry, D. 2009. "Two systems for empathy: A double dissociation between emotional and cognitive empathy in inferior frontal gyrus versus ventromedial prefrontal lesions" in *Brain* 132.3: 617–627.

Skoyles, J.R. 2008. "Why our brains cherish humanity: Mirror neurons and colamus humanitatem" in *Avances en Psicología Latinoamericana* 26.1: 99–111.

Sugita, Y. 2009. "Innate face processing" in *Current Opinion in Neurobiology* 19.1: 39–44.

Taplin, O. 1977. *The stagecraft of Aeschylus: The dramatic use of exits and entrances in Greek tragedy*. Oxford: Oxford University Press.

Trevarthen, C. and Fresquez, C. 2015. "Sharing human movement for well-being: Research on communication in infancy and applications in dance movement psychotherapy" in *Body, Movement and Dance in Psychotherapy* 10.4: 194–210.

Turner, V. 1979. "Frame, flow and reflection: Ritual and drama as public liminality" in *Japanese Journal of Religious Studies* 6: 465–499.

Valakas, K. 2002. "The use of the body by actors in tragedy and satyr play" in Easterling, P. and Hall, E. (Eds.), *Greek and Roman actors: Aspects of an ancient profession*. Cambridge: Cambridge University Press: 69–92.

Van den Stock, J., Righart, R. and de Gelder, B. 2007. "Body expressions influence recognition of emotions in the face and voice" in *Emotion* 7.3: 487–494.

Van Rysewyk, S. 2011. "Beyond faces: The relevance of Moebius syndrome to emotion recognition and empathy" in *Emotional Expression: The Brain and the Face* 3: 75–97.

Verbeke, W. 1997. "Individual differences in emotional contagion of salespersons: Its effect on performance and burnout" in *Psychology & Marketing* 14.6: 617–636.

Warden, C. 2012. *British avant-garde theatre*. London: Palgrave Macmillan.

Webb, R. 2008. "Inside the mask: Pantomime from the performers' Perspective" in Hall, E. and Wyles, R. (Eds.), *New directions in ancient pantomime*. Oxford: Oxford University Press: 43–60.

Wicker, B., Keysers, C., Plailly, J., Royet, J.-P., Gallese, V. and Rizzolatti, G. 2003. "Both of us disgusted in my insula: The common neural basis of seeing and feeling disgust" in *Neuron* 40.3: 655–664.

Wyles, R. 2011. *Costume in Greek tragedy*. Bristol, UK: Bristol Classical Press.

Zaki, J. and Ochsner, K. 2013. "Neural sources of empathy: An evolving story" in Baron-Cohen, S., Lombardo, M. and Tager-Flusberg, H. (Eds.), *Understanding other minds: Perspectives from developmental social neuroscience*. Oxford: Oxford University Press: 214–232.

Zieber, N., Kangas, A., Hock, A. and Bhatt, R.S. 2014. "Infants' perception of emotion from body movements" in *Child Development* 85.2: 675–684.

5   *Melos*

Music and the mind

Aristotle placed *lexis* or "language" fourth on his list of the elements of tragedy with "song/music making" (*melopoiia*) next. In these next two chapters I want to examine the words and music of Greek drama from the interlinked perspective of how they were received in performance as music *with* words. The words of ancient drama were received either as lyrics set to music or as a heightened form of verbal delivery following distinct metrical patterns with an inherent musicality of pitch variation and rhythm; therefore, in this chapter we will explore the music of drama from a cognitive perspective in order to try to better understand its effect on the audience and set a performative context for *lexis*. Aristotle describes *melopoiia* as the most important "embellishment" in his list of the parts of tragedy; the term Aristotle uses is *hedusma*, which also means "relish" or "seasoning," and this reminds us of the dramatic flavors of Indian *Rasa* (Poetics 1450b17–19). Although Rasa theory can show us the efficacy of such "flavors" when it comes to the totality of the performance form, I want to challenge Aristotle's view on music, that it is not just "auditory cheesecake," to paraphrase Steven Pinker,[1] but one of the most effective communicators of emotion available to the Greek dramatist and another effective way of creating cognitive absorption, attention, dissociation, and empathy.

In antiquity, Athenian drama was regarded as a musical form insofar as the performance of poetry was conveyed by music – so much so that Plato thought it could corrupt listener's souls, Aristotle thought it could inspire *catharsis*, and Plutarch wrote that Athenian prisoners of war could buy their freedom with it. In antiquity, the music of the Athenian theatre was variously described as "loud, bombastic and inaccessible" or "orgiastic"; it was said to cause "a frenzied and excessive lust for pleasure"; it was nothing but "effeminate twitterings"; it made "people go crazy for even the tiniest fragments"; and it had the power "to change the soul."[2] Both Plato and Aristotle railed against it, and the comedies of Aristophanes are full of parodies, jokes, arguments, and physical brawls over the "right" kind of theatrical music. There can be no doubt that the music of Greek drama had a massive impact. With that being said, we face a serious challenge. Only three tiny scraps of Greek tragic musical notation have survived, and as yet we still have no idea what this controversial and, evidently, hugely popular music actually sounded like.

According to a recent study by Pöhlman and West, there are some 61 fragments of Greek musical notation currently known, spanning a period of around 1,000 years.[3] Pöhlman and West have done a great deal to advance our knowledge of ancient Greek music and even enabled certain early songs to be transcribed into Western musical notation and played today. This work is based on the theory that the rise and fall of the pitch tones in the music is reflected in the cadence of the tonal qualities of the sung Greek. This is probably correct, as we see a distinct reaction in antiquity when music starts to diverge from the metrical and pitch qualities of the lyrics, but we cannot know if these reconstructions really sound like the music the Greeks listened to in antiquity. As I write, the classicist and musician Armand D'Angour is attempting to reconstruct the music of tragedy from the available evidence and ethnomusicological studies.[4] One can only hope that this project helps us to know Greek dramatic music better. In the meantime, we really have very little idea of how the music of tragedy actually sounded.

Though we can no longer hear the music of Greek drama, a cognitive model can help us to understand more about how it functioned in performance and was received by the audience it was originally intended for. This chapter starts by examining the cognitive attributes of the *aulos*, the twin-piped reed instrument that accompanied drama, and its superexpressive voice-like qualities. It then focuses on the music of tragedy, particularly the lament, which was one of its most ubiquitous musical forms. This poses the musical version of the tragic paradox: why and how did the Greeks want to be entertained by listening to highly emotional music? I propose that it fostered a sense of dissociation, which led to cognitive absorption and increased empathetic feelings. Finally, I examine music from a predictive perspective and the cognitive mechanisms of music expectancy in particular before surmising that one of the most powerful effects of tragic music was the creation of a dissociative and absorbing atmosphere within which empathetic feelings could be promoted.

Cross-cultural music

In many ways, music is one of the best cultural products with which to explore cross-cultural emotional and aesthetic responses. Biological psychologist Stefan Koelsch writes, "the importance of music for humans, as well as the capability of music to strongly affect emotions and mood in most humans, makes music an extremely interesting experimental stimulus for affective neuroscientists."[5] Koelsch has made the following important observations about the universality of music across human cultures:

- Humans make music in every known culture.
- Up until the advent of recording devices in the last 100 years, music was primarily a social event.
- Humans possess an inborn neural architecture for musical comprehension, sensitive to pitch, rhythm, consonance and dissonance at birth.

- Infant-directed singing of lullabies and play-songs is very similar across cultures.
- The expression of emotion in music can be discerned across cultures.

Although ancient Greek music may have sounded completely alien and even unappealing to our ears today, we share with the ancient Athenian the same basic physical neural mechanisms for processing music. Kathleen Higgins agrees with the cross-cultural connectivity of how humans process the emotional qualities of music and concludes:

> Music's rhythmic reflection of vitality, its ability to entrain its listeners, and its participatory enlistment of the body are all ways in which our biology is engaged in provoking affect. Culture fleshes out many of the details of the emotional experience, yet we have every reason to think that music can stir emotions of solidarity across cultural divides.[6]

Cross-cultural studies of musical processing are important for our purposes. If we can establish some basic commonalities across different time periods and distinct cultures in human musical processing, then there is scientific validity to applying this research to the ancient Greeks, as part of a wider bio-cultural methodology. One such influential study took place in 2009, led by Thomas Fritz. It compared the musical responses to emotional music between 21 members of the Mafa people of Cameroon and 20 German participants.[7] The Mafa were chosen because they have had little exposure to Western music, and their compositions, produced on a series of small horns, each with its own distinct pitch, sound quite alien to most Western listeners. The researchers played 42 pieces of Western classical, pop, jazz, and rock music, generally recognized as being "happy," "sad," or "frightening" to Mafa participants. Then Mafa musical pieces, given the same three emotional descriptions in Mafa culture, were played to the German participants.

The researchers were seeking to establish if "basic" emotions in music could transcend cultural influences and taste preferences.[8] What they found is that the participants from both cultural groups recognized the emotional qualities of the foreign music they heard. The researchers also electronically modulated the music, making it either more or less dissonant, by altering the harmony, pitch, and intervals. They found that both the Mafa and German participants had similar negative responses to dissonance in the music even when the music was unfamiliar and strange to them. This study concluded that while the extent of such musical modulation is clearly influenced by culture, consonance and sensory dissonance affects the perceived pleasantness of music across different cultures.

A similar study, also carried out in 2009, used 10 different excerpts of Byzantine music that had survived in the Greek musical tradition and tested 14 Italian, 15 British, and 31 Greek participants.[9] Again, each group was quite successful at identifying the emotional qualities of the Greek music, though it was unfamiliar to the Italian and British subjects. The difference was how each group classed the intensity of the emotions. This seems to be a facet of cultural exposure and learned associations held in the working memory. These studies are part of a growing body of research that have indicated we can make some basic assumptions about the way music is received and perceived across different cultural groups. This gives us some common ground to start to consider the music of Greek drama and the effect it had on its original audience.

Some universal biological elements can be applied. For example, humans possess the ability to preserve sound-generated information as it enters the auditory processing mechanisms and is processed in various areas of the brain. This is how we can hold on to the meaning of a long sentence until it reaches its conclusion or connect the different pitches of a lengthy strain of music and appreciate its melodic intricacies. This may be unique to humans, as there is no evidence that other primates can do this, despite having the ability to process complex visual signals. This ability, which seems to be innate, involves the deployment of working memory, first formed in the womb as the brain systems are developing, and then during infancy and on into childhood. In this way, we remember the music we grow up with, and its melodic, tonal, and rhythmic familiarity creates cognitive templates upon which we compare the incoming sounds we hear. This is why expectation is a key element in our ability to enjoy music.[10] Music also strongly provokes our emotional processing systems, with rhythm, pitch, and tone acting upon our autonomic bodily regulation systems to produce marked physical embodied responses.

There is a continuity of tonality of scale across different human cultures and time periods, and this is closely associated with the pitch range of the human voice. The Mafa study surmised that the participants were adept at recognizing the emotional qualities of completely unfamiliar music because their auditory processing systems were interpreting "tone of voice" qualities in the music. These are expressed by tempo, pitch, intervals, and harmonics. In this respect, music is a product of human distributed cognition: it is produced by human movement acting upon an external musical object, it is embodied by pitch scales derived from the human vocal range, and it is conveyed by bodily entrained rhythms and musical intervals based on mutually communicable human expectations. With this in mind, some instruments are particularly adept at replicating the same kind of affective states that react to sounds produced by the human vocal system, whether calming, alarming, terrifying, or sad. Such instruments have been called "super-expressive" in that they re-create and magnify the pitch qualities and emotional aspects of the human voice. The instrument that accompanied the lyrics of Greek drama, the aulos, was such a super-expressive instrument.

The super-expressive aulos

According to Juslin, certain instruments that are super-expressive, such as the violin, amplified guitar, flute, and oboe, act on the brain as extreme forms of the human voice and provoke heightened cognitive responses.[11] These instruments can outdo the voice in terms of speed, timbre, and intensity, and this can elicit powerful affective responses and even promote emotional contagion – the way in which the emotions of two or more individuals can converge, or be mirrored, both implicitly and explicitly.

Humans have listened to a super-expressive voice generated by a pipe that can change its pitch to create a melody for at least 40,000 years. In 2003 a bone flute was found in Hohle Fels in southwestern Germany and dated to the Upper Paleolithic period, from between 42,000 and 35,000 years old. This seemingly rudimentary instrument was made from the radius bone of a vulture and has five human-made finger holes and lines adjacent to them with one end carved in a notch where it was blown.[12] What is remarkable is that the holes on this Paleolithic flute are set out in line with a diatonic scale, where the octave is divided into five whole steps and two half steps. When a reconstruction of the ancient flute was created in 2004, it was possible to play the main theme of Bach's *The Art of the Fugue*.[13]

The aulos, the primary instrument used in Greek drama, may have been a much more sophisticated version of the Hohle Fels bone flute, but it functioned on a cognitive basis in the exact same way. The aulos was made up of two long pipes, which were blown through reeds. The reeds provided its penetrating "buzzing" sound at the lower registers and its so-called geese-like screeching at higher pitches.[14] The length of the pipe established its pitch range and volume, and finger holes could change its pitch by as much as an octave to create a melody. While we can still only speculate on the actual music that the aulos was used to play in classical dramatic performances, we can know how this distinctive instrument significantly heightened the affective states of its listeners. This was in large part due to the way its evocative sound acted on the mind's music-processing neural networks, and in turn how it greatly helped create a dissociative and dissonant aesthetic environment.

If the lyre was the primary musical instrument that accompanied epic and lyric, the aulos came to dominate in the theatre. Its penetrating tone and volume meant that it could be heard alongside the multiple voices of the chorus, and it could reach the back row of a *theatron* that accommodated around 6,000 people. The aulos could be heard over the sound of wind, marching soldiers, the creaking of a warship, or the din of spectators gathered at a public event. Thus, in Athenian culture, the aulos was used to drill troops, regulate the strokes of oarsmen, herald athletic events, and provide the musical accompaniment for dithyramb, tragedy, comedy, and satyr plays, which were all staged in large open-air spaces. The open-air theatre is predominantly a low-frequency aural environment, made up of wind noise, audience murmuring, and crowd movement. In the late fourth century, the stone seat risers of the theatre at Epidauros solved the problem of this ambient noise. These were baffled to absorb low-frequency sound and enhance the vocal and voice-like pitches.[15] Although the fifth-century Athenian theatre space was considerably smaller, it was still a large-size open-air venue, and the super-expressive voice-like qualities of the aulos and its volume-producing attributes were perfectly suited for such an environment.

The aulos was frequently associated with strong emotional responses in antiquity: Plato regarded the instrument as having the power to overtake the listener with frenzy and induce religious purification,[16] and Aristotle described it as "not a moral instrument but rather one that excites the emotions."[17] In the Athenian theatre of the fifth century, the penetrating tones of the aulos frequently played foreign "Phrygian" strains and was used to accompany the performance of lamentations for the dead. Peter Wilson has described it as "a danger: it threatened self-control; it marred the aesthetics of the body; it introduced the allure of the alien," and Andrew Barker has written that "the music of the aulos was dramatic and emotional; it was versatile in mood and effect. Capable of blaring vigour, plangent lamentation or sensual suggestiveness."[18]

Despite its ubiquity in Athenian performance culture, the aulos was regarded as strange, foreign, exotic, and dangerous, and its expert players were never Athenians but outsiders from other cities. With that being said, as the music of dithyramb and tragedy in particular became more complex and virtuosic, these foreign aulos players became increasingly admired for their skill and often became famous, commanding large performance fees, and they were in high demand across the Greek world.[19] The Pronomos vase (Chapter Three, Figure 3.1) is named for the aulos player, and he sits in pride of place directly underneath Dionysos. In classical Athens, aulos players were the exotic rock stars of their day.

The aulos was well known as a highly effective communicator of emotional range, yet this aspect of the instrument caused it to be frowned upon by many elite Athenians. This attitude is reflected in Plato's *Republic*, where only "the instruments of Apollo" (lyres) would be admitted to the ideal city and not the aulos (399d-e). Also, Aristotle regarded the aulos with particular contempt, as it substituted the voice with sounds that he described as provoking "orgiastic influence" and better suited for "cathartic theoria" (purification spectacles, such as the theatre) than for moral instruction (*Politics* 1341a). Plutarch's Alcibiades also rejects the aulos because it distorted his face as he puffed out his cheeks to play it and "blocked up and barricaded the mouth, robbing its master of both voice and speech."[20] This anecdote reflects a mythic tradition that told how Athena rejected the aulos almost as soon as she created it and threw it away after she saw how her cheeks puffed up and made her look unsightly. A satyr named Marsyas tried to recover the pipes for himself, and he was struck by Athena for daring to be so presumptuous. These stories encapsulate the subversive and dissociative qualities of the aulos in Greek culture and how it replaced the human voice and distorted the player's visage.

The ability of the aulos to act as a super-expressive voice imbued it with such a reputation for emotionality in antiquity.[21] The Greeks were well aware of the voice-like quality of this instrument: it is frequently described as "many-voiced," called an "extraneous voice" or even "sweet-voiced."[22] Pindar associated its sound with intense cries of terror and pain and relates a story of how after Perseus had beheaded the Gorgon, Medusa, Athena created the "many-voiced" song of the aulos so she could imitate the "shrill cry that reached her ears from the fast-moving Gorgon." Athena then names the aulos "many-headed," and its sound is "the famous enticing call for people to come to festivals" (*Pythian* 12). In the same vein, Aristotle understood the power of the aulos to excite emotions and linked it with the Phrygian mode of music – both elicited ecstasy and emotion. He was also aware of its voice-like qualities and pointed out that lyrics accompanied by the aulos are far more pleasant than those accompanied by the lyre:

> The song and the aulos mingle with one another because of their similarity, since both arise through the breath, while the note of the lyre, either because it does not arise through the breath or because it is less perceptible than the sound of the *auloi*, is less capable of mixing with the voice because it creates a distinction in perception.
>
> Aristotle, *Politics* 1342b,1–5

The vocal properties of the aulos can also be observed in its construction. In a survey of the many images of the aulos found on vase paintings and sculpture in the classical period, Stefan Hagel has deduced that the average iconographic aulos measures 44.5 cm from the player's mouth. This research places pitch ranges at over an octave in difference but shows a distinct pitch concentration of around 185 Hz, the median frequency for the human voice, falling into the contralto/alto range.[23] Ancient commentaries on the types of auloi also emphasize the vocal similarities. Aristoxenus names five types of auloi: *parthenioi, paidikoi, kitharisterioi, teleioi*, and *hyperteleioi*. These names translate as "girl-like," "boy-like," "lyre playing-like," "adult-type," and "super-adult-type." Most reflect human vocal pitches, with the exception of the lyre-playing-like aulos, which we assume was used to accompany the stringed lyre.[24] Likewise, in his *Onomasticon* (4.81) Pollux associates the *parthenoi* with girls, the *paidikoi* with boys, and the *hyperteleioi* with men and names the *teleioi* as an instrumental aulos. Herodotus also alludes to this classification of the aulos types equated with the human voice when he describes a Lydian army marching to the sound of "women's and men's auloi" (1.17.1). Thus, in antiquity we find the aulos frequently given the epithet *pamphonos* ("many-voiced").[25]

The strange, otherworldly sound of the aulos may have been doing more than activating the part of the auditory system that processes the human voice. Brain imaging research with fMRI has indicated that specific parts of the brain respond to human voices, and EEG studies have shown that auditory processing is significantly faster for vocally generated sounds than for non-vocal sounds.[26] Furthermore, evidence from research carried out on people with brain injuries indicates that humans have separate neural networks that differentiate between emotional vocal sounds and emotional musical sounds. What a super-expressive voice, like the aulos, may be doing is sending both of these auditory systems into a cognitively demanding and emotionally productive state, creating an intense affective experience for the listener. Such musical effects can manifest themselves in physical form, by changing cardiovascular and ventilation rates, temperature regulation, and hormone production. We use embodied language to describe what we often feel at these musically induced moments: we feel "chills" or events are "spine-tingling"; we get "goosebumps" or "lose our breath"; music can "make the hair on the back of my neck stand up" and induce these "frissons" or other autonomic affective sensations.[27]

Consider the use of background music in film. It often rumbles along at a low-frequency register – we are aware of it, underscoring a scene, but it is not the primary perceptive element we might be paying attention to. We might recall John Williams' famous score for *Jaws* (1975). When the film wishes to communicate a sense of subliminal unease, we hear the low entrained throb of super-expressive violin strings gaining in rate and intensity, like the pulse of a quickening heart rate. The listener's breathing and heart rate increases accordingly, skin temperature changes, and the chest can even tighten, adding to a sense of tension and expectancy. But when the shark suddenly attacks, the music shifts to the harsh stabbing timbre of high-frequency brass sounds – a super-expressive voice reminiscent of human screams of terror and pain. It's still quite difficult to hear that soundtrack and not feel more than a little stressed out. Another super-expressive voice is the music for the shower scene in Hitchcock's *Psycho* (1960) by Bernard Hermann. Here, the familiar and calming low-frequency sound of the running water of a shower is jarringly punctuated by the screeching of piercing violins, which are then joined by the sound of real human screaming. It's the sound of those stabbing super-expressive violins that prevented a generation from ever feeling safe in the shower again.

The music of Greek drama and the lyrics that were sung to it were frequently described as highly emotional, excitable, and even disruptive, all of which were epithets also associated with the aulos. For example, Aristotle links the aulos and the exotic Phrygian mode of music with the performance of "poetry and all Bacchic performances and movement" (by which he means theatrical music) and that both are "excitable and emotional" (*Poetics* 8.1342b 1–3). In this respect, the aulos was the most Dionysian of musical instruments, and it was quite appropriate that it accompanied the dithyramb, tragedy, and comedy. In Sophocles' *Women of Trachis*, the chorus of women rejoices at the news that Herakles is returning home to his wife alive and well:

> *I am exalted! I will not*
> *Spurn the aulos*
> *You are the master of my mind*
> *Look, his ivy drives me into frenzy!*
> *Euoi!*
> *I whirl round in the grip of Bacchus*
> *Io! Io! Paian*
>
> Sophocles, *Women of*
> *Trachis* 216–220

The rites of Dionysos, based on wine, music, and dance, were envisioned as accompanied by the super-expressive vocal sounds of the aulos. This could produce a kind of temporary insanity. As the chorus of Sophocles' *Women of Trachis* give themselves over to Dionysos, they begin to make their own super-expressive, seemingly non-sensical, vocal sounds: "Euoi!" "Io! Io!" "Paian!" The links between Dionysos and these kind of emotional, unintelligible cries will be explored at the end of this chapter.

In Greek drama music did not operate in isolation and was part of a total theatre work that incorporated festival, space, mask, movement, plot, language, and action. Yet above it all rose the sound of the aulos – strange, haunting, frenetic, and beguiling. It was an absorbing, dissonant sound, two important cognitive elements central to empathy that I will return to more fully in the next chapter. In the meantime, to appreciate how the strains of the aulos were interwoven into the soundscape of Greek drama, let us focus on one of the most prevalent musical genres of tragedy, the lament, and probe the question: why would such a strange, foreign-sounding instrument playing music associated with loss and death become so wildly popular – to the point that both Plato and Aristotle wanted it regulated or even banned? Perhaps because the music of tragedy could drive its audience theatre-mad.

The dissonant lament

The Syrian/Roman writer Lucian begins his treatise, *How to Write History*, with a remarkable tale:

> There is a story of a strange epidemic at Abdera, just after King Lysimachus took power. It began with the whole population exhibiting pronounced fever-ish symptoms, which were unrelenting from the very first attack. Around the seventh day, the fever was broken, in some cases by a violent nosebleed, in others by aggressive perspiration. Most absurd was their emotional state of mind. They were all beside themselves with tragedy and screaming iambics at the top of their voices. They loved to sing monodies from the *Andromeda* of Euripides and one after another they would recite parts of the speech of Perseus. The whole city was full of pale husks of men – their own seven-day tragic chorus singing;
>
> *O Love, who rules both Gods and men!*

Lucian gives the reason for this incredible plague as a visit by the famous tragic actor Archelaus, during the height of summer, to a city that perhaps had never before seen theatre. This anecdote is no doubt an exaggeration, but it is based on the real impact of ancient drama as it spread throughout the Greek world[28] and conveys the physical and emotional force that the theatre had on people who had never before experienced this revolutionary new art form. Perhaps what is most interesting about this story for our purposes is how the theatre-mad Abderans expressed their insanity by screaming at the top of their voices, vocalizing and singing. The performance and its music had affected them to the point that they had become "pale husks of men" – the embodiment of tragic waste.

The Abderans wandered around belting out laments until the cold weather came and provided a cure. This comic story shows us the power of the theatrical lament in antiquity, and as much as it made the Abderans temporarily insane it also incensed Plato, who attacked dramatic lamentations and cited his "chief accusation" against tragedy as its "power to corrupt." Plato wrote that when a performer is heard:

> (I)mitating one of the heroes sorrowing and making a long lamenting speech or singing and beating his breast, you know that we enjoy it, give ourselves up to accompanying it, empathizing with the hero, taking his sufferings seriously, and praise as a good poet the one who affects us most in this way . . . whereas when something afflicts us in real life, the opposite is true and we pride ourselves on our ability to remain calm and endure.
>
> Plato, *Republic* 10.605c-e

Peter Wilson has stressed that the aulos is "frequently associated with the sound of lamentation and many of its formal musical expressions" and that the use of the aulos, as the dominant musical instrument of tragedy, was "vital to a genre where the music of lament is the dominant form."[29] Lamentation, or the funeral song, was the most pervasive musical form in Greek tragedy: in the corpus of extant complete Greek tragedies (not the many fragments) there are 42 identifiable laments: 26 sung by women and 18 by men.[30] The three papyrus scraps of tragic musical notation that we have are all lamentations: the Vienna Papyrus begins with the refrain "I Lament, I lament"; the lines quoted by Dionysius (*Comp.* 63f) from *Orestes* are also in the form of a mourning song; and the two parts of *Iphigenia at Aulis* on the Leiden Papyrus are a choral lamentation and a *kommos* between Iphigenia and the Chorus.

In fifth-century Athens, public lamentation rites were socially controversial and perceived as female and foreign, with no rightful place in the male-dominated democratic polis. Even though there are numerous staged lamentations in Greek drama, we cannot know if these depicted actual contemporary funerary practice.[31] A good number of images of lamentation and mourning are shown on vase paintings and sculptural monuments, as well as accounts of funerary practice from non-dramatic literary sources. From these we can tell that Greek funerals were collective events that involved the preparing of the dead for burial, the laying out of the body, and ritual mourning by family members and in some cases the wider community. The act of mourning was a public display and involved the singing of funeral laments and choreographed ritual movements, such as the striking of the head, tearing of clothing, and extreme gestures of grief. In the years before the organization of the state-run theatre festivals, this public part of the funeral ritual became somewhat professionalized, with wealthy families hiring choruses of mourners to join the lament. In this way, the aristocratic funeral became a means of publicly signifying status and securing the continuity of power within the wider community.

Since the sixth century BCE, there had been successive legislation in Greece (not just Athens) limiting these public displays of lamentation. This may have been politically motivated to curtail the influence of aristocratic families. Limits were placed on the length of time that could be devoted to mourning; funeral processions had to be held before sunrise; and the amount of clothing buried with the corpse was restricted, as was the amount of time that could be spent erecting a tomb or grave marker. Although both men and women practiced public lamentation, the expression of excessive emotion started to be associated with female funerary rites and frowned upon by the men in power.[32] This may be yet another reason for Plato's negativity toward the music of lamentation: he associated it with the "irrational acts" of women. In the male-dominated democracy of fifth-century Athens, the female voice was progressively silenced, by both legislation and contemporary cultural practice.

The Athenian democracy was defined as the "power" (*kratos*) of the citizen men of the "districts" (*demos*). In this social structure the male-led state (*polis*) was supreme, not the female sphere of the household (*oikos*). Athenian social politics in the fifth century, and much of the content of drama, reflected this tension, which also played out between aristocratic and democratic factions in the city. Although Athenian plays were a product of democratic Athens, they were financed by aristocrats, who, if no longer able to publicly display their family's prestige by means of a large public funeral, could do so by producing a play. It is notable then that many of these plays contained scenes of fictionalized lamentation, something that was, by now, rarely seen in Athenian public culture.

The place of music and lamentation in the performance and reception of tragedy has long fascinated scholars of ancient drama. In the early twentieth century, Gilbert Murray, influenced by the anthropological Cambridge School, proposed a ritual structure to Greek tragedy derived from cyclical harvest festivals.[33] The performance of lamentation was an important element in their theory that tragedy developed from the performance of earlier ritual forms connected with the seasons and human rites of passage. Murray's ideas on drama and his new translations were embraced by the public at this time, as productions of Greek plays began to proliferate in universities and on the professional stage in Great Britain and the United States.

In the 1930s and 40s, Swedish classicist Martin Nilsson moved away from the seasonal cycle theories of Murray but still suggested that tragedy developed directly from ritual lamentation and hero-cult.[34] Following Nilsson, the American scholar Gerald Else proposed that Tragedy was the invention of Thespis and that lamentation was an essential element.[35] Even the anti-ritualist, Sir Arthur Pickard-Cambridge, who produced an influential thorough examination of the literary and epigraphic evidence for Greek drama, pointed out that the dithyramb, which he proposed tragedy grew from, was accompanied by the aulos and had a reputation in antiquity for being highly emotional.[36]

More recently, Margaret Alexiou has posited that there is an unbroken tradition of women's lament from antiquity to present-day Greece and that the women preserve traditional forms of lamentation in many cultures. According to Alexiou, although the lamentations staged in Greek drama may not have been faithful reconstructions of contemporary practice, they nevertheless would have strongly reminded the audience of the traditional songs of their mothers and grandmothers.[37] Subsequently, Nicole Loraux proposed that from the perspective of Athenian state theatre, theatrical lamentation was a "re-presentation" of real ritual practice and not something that could be watched only intellectually or aesthetically. Loraux advanced a compelling theory that tragedy was acting as a force for social amnesty: "the outpouring of grief allowed by the singing of the lament acts as a means of dispelling feelings of unresolved grief that can easily spill over into rage and revenge."[38]

The lament was not the only form of music heard in drama; there were also victory songs, prayers, wedding songs, and other traditional Greek musical forms. All of these genres would have evoked common cultural memories and greatly enhanced emotional contagion amongst the audience. In tragedy a good deal of this music is performed as part of the broader narrative of the play, often in an incongruous and dissociative manner.

One fine example of this in practice occurs in Aeschylus's *Persians* (249–510): a Persian (and therefore enemy) messenger has returned to Susa to tell of the Persian defeat at Salamis. His news devastates the chorus of Persian elders. During his speech the messenger evokes the spirit of the Athenian rowers by singing their Paean – the song they sung to steel their spirits as they rowed furiously to join the great sea battle. Bearing in mind that *Persians* was staged a mere eight years after the events being described, one can surmise that the majority of the audience not only knew this Paean well but many of them may have been the ones singing it at the actual battle in 480. Yet, here Aeschylus has the song performed by a defeated Persian. This must have led to complex feelings amongst the Athenian audience members – the terror of the attack, the chaos of battle, the trauma of the carnage, their own national pride, religious sentiments, as the song was offered to the gods, and a mixture of feelings about the rendition here by the Persian messenger. What this example exemplifies is how the lost music of Greek drama was one of the most effective and powerful communicators of emotional contagion.

A musical tragic paradox

Plato wrote that people felt pleasure when watching lamentation and abandoned themselves to it (*Republic* 605c), and that they derived both pleasure and pain from songs of mourning (*Philebus* 48a). Why did the Greeks derive so much pleasure from listening to sad or disturbing music? For example, Xenophon wrote about how the actor Callipides could fill a theatre, because he could move the entire audience to tears (*Symposium* 3.11). This penchant for seeking pleasure in sad music was not confined to the Athenians: Pausanias informs us that the only men who returned from the disastrous Sicilian Expedition alive were those who could buy their freedom by entertaining the Greek colonists of Sicily with the tragic songs of Euripides.[39]

People across many diverse cultural and ethnic groups share this tendency to derive pleasure from highly emotional music. In the West this might be our enjoyment of opera, blues, rock ballads, classical requiems, or weepy pop songs. Cognitive musicologists and those working in the affective sciences are very interested in several questions related to the popularity of sad music, such as, when we listen to sad music do we also become sad? How does processing sad music affect our neurochemistry? Does listening to sad music have an effect on our cognitive reward systems? These questions are in effect a musical version of Hume's tragic paradox: the emotional conundrum that many people seemingly derive pleasure from the representation of other people's grief.[40]

Of course, different people will respond differently to emotional music depending on a wide variety of biological, contextual, and cultural factors.[41] Yet, despite this seemingly obvious assertion, most of us have experienced those incredible moments at a live performance when the entire audience seems caught up in the same mimetic event. It is as if everyone is breathing together and intently focused on the same moment: they are "on the edge of their seats." Performance artists crave these moments – when the audience becomes fully absorbed and seems to be responding in unison, as a collective whole, and an emotional *frisson* permeates the theatre space.[42] Like the moments of attention that are produced by the error corrections of predictive processing, so the dissociative elements of the ancient stage, with music being a key conveyer, could foster a collective cognitive absorption.

Listening to music expressing a set of emotional values while feeling another has also been connected to mind-body dissonance, which has been shown to increase judgment accuracy and empathy.[43] We will delve into the cognitive effects of the kind of dissociation created by Greek dramatic music, masks, movement, narratives, and environment in Chapter Seven, but in the meantime it will be useful to analyze what we can ascertain about the effects of tragic music within the theoretical framework of predictive processing. We have seen how the kind of emotionally affecting dissonant sounds generated by dramatic lamentations can act very quickly on our sensory processing systems and force a rapid embodied response. These have been called brainstem reflexes, and they can seem autonomic.[44] In musical terms such sounds might be sudden cries, loud noises, dissonant pitch, changing tempos, a shrill note, a sudden shift in meter, or an unfamiliar melody – all are common features of lament. The aulos was also quite adept at eliciting brainstem reflexive responses with its ability to produce piercing, sudden, and strange sounds. This leads to the question as to whether these kinds of sounds are exteroceptive – in that they are generated by an exterior auditory source – or interoceptive – in that the embodied responses generated by brainstem reflexes also add to our multi-modal arsenal of predictive information gathering. The visceral embodied feelings that such sounds can produce have been recently analyzed within the framework of Friston's free energy principle – the "concurrent dynamical updating of expectations about the causes of external (exteroceptive) and internal (interoceptive and proprioceptive) sensory inputs."[45] Within this model, emotion is viewed as the active inference of both internal and external bodily sensations.[46] This is similar to what Jesse Prinz has called the embodiment theory of emotions.[47] Ondobaka, Kilner, and Friston hypothesize:

> Interoception, formulated under active inference, plays a fundamental role in ToM [Theory of Mind]. From the active inference perspective, knowing

the contents of another's mind can be demystified and simply recast as an optimal explanation for perceived (motor and visceral) behaviour in others – that would have been produced by ourselves, have we had been in the same intentional and emotional state.[48]

Watching and understanding drama is a mimetic version of Theory of Mind, where we are expected as audience members to try to understand and predict the motivation of the characters before us. Making inferences about the minds of others is a type of "interoceptive mirroring system." Whereas exteroceptive and proprioceptive inference are "on view," it is much harder to gain an understanding of another's internal visceral state, but we learn to detect clues from our own social interactions. This is not to say that we always "feel" the same emotion as the mind we are making inferences about – a prediction of anger in another can result in the emotion of fear in us; however, interoceptive inferences based on our own embodied responses work in tandem with our other perceptual generators, but they are far more deeply "felt" because of the rapid and autonomic nature of our embodied visceral responses. Hence, the power of unintelligible dissonant sounds, which will be discussed briefly below, the seemingly changing visage of the mask, and the kinesthetic empathy generated by watching movement and, of course, listening to music.

The performance of lamentation, at once familiar, recalling old familial rituals, and strange, with dissonant tones, big open vowel sounds, super-expressive instrumentation, entrained movements, and disturbing lyrics, was a highly visceral experience. These interoceptive inferences promoted a Theory of Mind for the characters, who although at first might have seemed remote and other, are now understood and empathized with, through the sharing of the music, movements, and lyrics of grief. By embodying the mimetic grief of fictional characters who were often very removed from the audience's own cultural and social experience, the audience members, by interoceptive inference, could start to feel their own grief, which could have been magnified by the emotional contagion generated by other audience members. The theatre then became a place for the expression of grief, an outlet for emotional expression, an affective mirror for a kind of cultural therapy within a society traumatized by conflict and war – what Aristotle called *catharsis*.[49] Perhaps this provides an answer to Loraux's question: "what happens to passion in the city-state?"[50]

Musical dissonance and absorption

The aulos could both beguile and unsettle, and its super-expressive voice contributed largely to the aesthetic and affective atmosphere of the classical Greek theatre. In addition to its emotional voice-like qualities, the aulos made a strange, otherworldly sound and was frequently used to play musical styles that were regarded as strange and foreign. In addition, the lament was prominent in tragedy, and its sounds ranged from mournful and distressing to wild and highly emotional. Emery Schubert has proposed that negative music heard in certain aesthetic contexts, such as drama, triggers what he terms a "dissociation node," which inhibits the displeasure circuits of the brain.[51] Dissociation is defined as the functional alteration of memory, perception, and identity and can occur in response to stress or trauma; within social rituals, such as traditional healing practices; and in response to aesthetic performances – in some extreme cases, spontaneously.[52] People have differing degrees of dissociation depending on a variety of social, cultural, and biological factors, but those who are more prone to dissociative experiences leading to cognitive absorption show a higher capacity to enjoy negative emotions in music.[53]

Dissociation also involves a cognitive disconnection from traumatic experiences and pain. Garrido and Schubert suggest that this same psychological process happens when listening to sad or strange music. The most common form of dissociation is absorption, which they have described as "the heart of the normative dissociative process."[54] Absorption is a state of focused immersion involving a narrowing of concentration and attention. It can include temporal dissociation (you lose track of time or become totally absorbed in a fictitious temporal pattern presented on stage) and often a sense of heightened pleasure. Self-concerns and the external world beyond the absorbing experience become irrelevant and critical thought is curtailed. Cognitive absorption is associated with aesthetic pleasure but also the state of mind one might experience when being mentally "consumed." This could be by a computer game or software program that is easy to use, useful, somewhat novel, and aesthetically pleasing.[55] Being "absorbed" by a piece of music, a play, or a film is reported by most people as being highly pleasurable.

For all of Plato's negativity toward tragic music, he seems to be aware of the importance of theatrical absorption. In *Laws* (665c) the Athenian remarks that it is the duty of every person and of the state to sing (or chant – the word has the sense of a charm or incantation) for each other and that these songs should "constantly change" and be "of variety in every way possible." This will in turn inspire the singers "with an insatiable appetite for music and the pleasure it provides." This sounds a lot like a description of cognitive absorption, especially the remark that the songs must constantly change – novelty being an important part of the process of aesthetic judgment. The Athenian has been describing choruses to Apollo and Dionysos, so Plato must have the theatre in mind here. To confirm this, he continues the dialogue with a discussion of how to encourage older men to not "feel ashamed" of getting up and singing in the theatre and notes that as men grow older they become more reluctant to sing songs and take less pleasure in doing so [665e].

Plato's Athenian suggest that the men over age 40 overcome this sense of shame by invoking Dionysos "above all gods" and inviting his presence at the rite, which the god gave to mortals as a medicine "against the crabbiness of old-age." This is so the old men might "renew their youth" through "forgetfulness of care" (dissociation). Plato wants the state to allow them to be assisted by an age-old means to an altered mental state: alcohol, and he proposes that at the festival of Dionysos they should be permitted, under the law, to get riotously drunk. The archaic poet Archilochus is credited with a similar sentiment when he famously sung, "I know how to lead off the dithryramb, my mind blasted with wine!"[56]

What the Dionysian wine helps bring is the intoxication of absorption. This is the same kind of altered mental state that is more gently enhanced by dopamine and more vigorously by music and dance. These elements – music, dance, masking, wine, the open air –constitute important elements in the worship of Dionysos. Hence in Euripides' *Bacchae*, the god makes no distinction between young and old and encourages them all to join him in the revels. The elderly Cadmus and the old prophet Tiresias are depicted dressed in fawn skins eagerly dancing as best they can. (*Bacchae* 200–209). Tiresias says that Bacchic revelry is a kind of madness that leads to "prophetic skill," as if the dissociative state of the Dionysian reveler leads to a kind of enlightenment. The chorus sings that Dionysos wants his followers:

> To join the dances
> To laugh with the *aulos*
> To bring an end to thought.
> Euripides, *Bacchae*
> 380-381

Plato was right that adults need an aid to reach a state of cognitive absorption. Research has shown that younger people do have higher rates of absorption.[57] Adolescents and young adults seem to function perfectly well with a high level of dissociative experiences in their lives without reporting that these are disruptive in any way. Plato seems to be suggesting that dissociative experiences (here performing in a chorus at a theatrical festival) should be part of everyday life in a healthy state. A body of contemporary research supports this claim: in the right context, cognitive shifts in consciousness that temporarily disable self-awareness and produce cognitive absorption seem to actually contribute to better functioning in other contexts and activities.[58] Absorption is different from our normal goal-driven cognitive functions and enables closer self-examination and deeper thinking about situations we might never have even imagined without the stimulus provided by a ritual, a performance, or an image. Absorption enables the creation of mental representations and, importantly, promotes feelings of empathetic recognition. We will return to the cognitive aspects of absorption in Chapter Seven and how the total effect of Greek drama promoted the mechanism of dissonance/dissociation, absorption, and empathy. To conclude this chapter on music, I wish to examine another important affective element that we can discern in ancient dramatic music: the development of musical expectancy, its affective attributes, and its relationship to prediction.

Prediction and musical expectancy

Music has a distinct syntactical structure based on pitch, melody, intervals, and rhythm. These forms become familiar to us within our respective cultures, and in this way music has a language or a grammar that we respond to. Our expectation of what the music is going to do is one of its greatest pleasures, and we can all think of a number of songs or pieces of music where we wait to take particular delight in a guitar solo, leitmotif, aria, or chorus. In fact, the structure of most modern pop songs exploits our capacity for musical expectancy, teaching the listener the pattern of the song as it is played with a variation of a simple formula comprising verse | verse | chorus |verse | chorus | chorus. I know I can't help listening to the Beatles' "Hey Jude" without the expectation of the final chorus, which is a superb mix of the unintelligible soaring vocal melody (*naa, naa, naa, na-na-na-naa, na-na-na-naa, Hey Jude*) – for me this music is truly infectious.

The concept of musical expectancy maps very well onto the idea of prediction as a fundamental human cognitive process, as Gebauer, Kringelbach, and Vuurst write:

> [M]usical anticipation and incongruity, i.e. elements that do not fit with schematic, veridical or short term memory-based predictions, may be a fundamental source of music emotion and pleasure. Hence, musical emotion and pleasure are driven by the dynamic interplay between the listener's expectations and the statistical regularities in the musical structure. Music is pleasurable when expectations are fulfilled, but probably even more so when they are slightly violated. According to this idea, unanticipated events are responded to with surprise, i.e. increased physiological arousal and optimized attention, but can be modulated by secondary cognitive appraisals of the event. So, the delights we get from unanticipated events in music are due to the contrast between our predictions and the musical structure, as well as the subsequent resolution within the music.[59]

Before we proceed further, it is well worth setting out the breakdown of the Bayesian predictive coding theory in the explanation by Gebauer, Kringelbach, and Vuurst:

Prediction	the process of minimizing prediction errors between higher-level "prediction units" and lower-level "error units."
Action	the active engagement of the motor system to resample the environment in order to reduce prediction error.
Emotion	a weight or modulator of the prediction error itself, guiding behavior, action and learning through neurotransmitters such as dopamine.
Learning	the long-term influence on the prediction units.

They conclude that music provides a dynamical interplay among perception, action and learning, and emotion, and it is therefore a constructed form of predictive processing.

When music delays our expectation – or even violates it – we notice very quickly, and this has an emotional effect. We can feel like we are "left hanging" or are taken along "a different path," and it is notable how much of the language we use to describe music is rooted in our embodied responses. A good many of the aesthetic opinions we find of music in antiquity are concerned with matters of musical expectancy and, as we will see, divergences in existing musical forms and structures were often met with confusion, derision, and outright anger. Aristophanes' comedy *Frogs* was produced at a time of crisis for the Athenians, around 405 BCE when civil unrest, revolution, and military defeats abounded. The premise of the play is that the city can be saved if only a great tragic poet can be resurrected to inspire the citizens to greatness. In the underworld, the comic Dionysos must decide between the recently deceased Euripides, well known for his musical innovations, and the long-dead and by now traditional poet Aeschylus. Many of their aesthetic objections to each other's works reflect an ancient fascination with musical expectancy. For example, Euripides complains at being frustrated for being made to wait for an Aeschylean veiled character silently mourning to begin to vocalize on stage. He goes on to condemn it as "a trick designed to keep spectators in their seats, waiting for when Niobe might start to speak" (919–920). Later, he describes Aeschylus's tragic music as monotonous "rope-makers' songs" (1296). Aeschylus counters that Euripides' musical phrasings can all be completed with the exact same metrical formula. But Aeschylus also attacks Euripides over his "indulgent" use of music, mocking the way he elongates words across the music, rather than having the music follow the tonal rise and fall of the spoken or sung Greek.[60]

> *Spiders in the corners of the roof*
> *Wi-yi-yi-yi-ind with your fingers*
> *The loom stretched thread.*
> 1309–1311

In Greek the elongated "wind" is "*eieieieilissete*" indicating that Aeschylus is parodying Euripides technique of stretching a word over several harmonic notes, something that is very familiar to listeners of contemporary music today. This vocal technique still has an air of artistic rebellion, non-conformity, and emotionality about it. If I may be permitted to commit a form of literary heresy, we can compare Euripides' technique with an example from 1980s pop group Duran Duran and their song *The Reflex*. The lyrics look like this:

> *So why don't you use it?*
> *Try not to bruise it*
> *Buy time don't lose it*
> *The reflex is an only child he's waiting in the park.*
> *The reflex is in charge of finding treasure in the dark.*

Those of us who admit to knowing the song remember that lead singer Simon Le Bon would stretch out the word "why" over several syllables to fit the melody, singing something like "Why-yi-yi-yi-yi don't you use it?" This technique is common in blues, rock and roll, soul music, pop, folk, and opera, but to some Athenians in the late fifth century, it was an absolute scandal as it defied their cultural musical expectations. It certainly incensed Plato, who had his "Athenian" make this complaint in *Laws* (700b-701a):

> Athenian music comprised various categories and forms (hymns, laments, paeans, dithyramb, lyre music) . . . once these categories and a number of others had been fixed, no one was allowed to pervert them by using one sort of tune in a composition belonging to another category. . . . Later, as time went on, composers arose who started to set a fashion of breaking the rules and offending good taste. They did have a natural artistic talent, but were ignorant of the correct and legitimate standards laid down by the Muse. Gripped by a frenzied and excessive lust for pleasure, they jumbled together laments and hymns, mixed paeans and dithyrambs and even imitated aulos tunes on the lyre.[61]

Ultimately, Plato's "Athenian" fears that musical lawlessness will lead to social anarchy, and he describes this new undisciplined society as a vicious *theatrocracy* (rule of the spectators) where pleasure is the only guiding artistic principle and every audience member is entitled to his own opinion. These new musical techniques were dramatized by Aristophanes in *Thesmophoriazusae*, where he has Euripides and his uncle visit the avant-garde dramatist Agathon. As he is revealed in the midst of his creative process, his servant describes what is happening:

> Agathon of the beautiful verses, our master, is about to . . . construct a frame for the hull of a play. He is bending new curves for his verses, chiseling the parts, gluing his songs together, mixing up maxims, fashioning paraphrases, making molds, smoothing, casting. . .
>
> Aristophanes, *Thesmophoiazusae* 39–69

It seems very familiar to us to read about people carping about the music they don't like; one can think of many modern examples, such as the audience response to Stravinsky's *Rite of Spring* in 1913, the reaction of middle-class America to Elvis Presley in the 1950s, Malcolm McClaren's exploitation of disaffected British youth to promote the *Sex Pistols* and punk rock – an idea exported from New York City, and the parental advisory stickers attached to the records of hip-hop and heavy metal artists.[62] Euripides had incorporated elements of what we call the "New Music" into his work, a musical revolution that started in the dithyrambs, celebratory songs known for their musical "twists and turns,"[63] in the mid to late fifth century as musicians became more professionalized and proficient in creating and executing different musical styles and qualities into their performances.

The development of the "New Music" from the performance tradition of the dithyramb and drama shows how ideas about musical expectancy carried real cultural and political force in Athenian society. Although we cannot hear the "New Music's" impact on tragedy, we can see its influence on the texts, such as the example from *Frogs* and in the later plays of Euripides, and we can discern what classicist Eric Csapo calls the effect of "sound over sense" in that the words are now responding to the music, not the other way around – as Plato would have it – with the music subservient to the text. These plays feature long, winding sentences and quite complicated aural effects such as alliteration, assonance, and repetition, especially at times of high emotion.[64] Choral passages are lessened, and there are far more solo arias by single actors. We also see the sense of the choral songs becoming more abstract and prone to visual imagery over narrative sense.

Most of the literary references we find to the "New Music" seem to be in response to having particular aspects of culturally conditioned musical expectancy disrupted. We hear of the tragic poet Agathon's meandering "ant trails" in Aristophanes, the dithyrambs composed by Timotheus are described as "perverted ant crawlings," and the musical innovator Philoxenus was actually named "The Ant."[65] Aristophanes was well aware of the cultural and political differences of contemporary Athenian attitudes to musical expectancy. In his comedy *Clouds*, he presents two characters in a verbal duel: the "Superior Argument," who purports to represent older, conservative Athenian values, and the "Inferior Argument," who stands for the new learning of Socrates and the teachers of rhetoric. The stuffy "Superior Argument" rants against the state of musical education and remembers fondly how things were in his day:

> And if any of them fooled around with the tune or twisted any twirls – the sort of knotted up twist we get these days from Phrynis – he was soundly beaten for obliterating the muses.
>
> Aristophanes, *Clouds* 969–972

The structure of music is an incredibly important cultural, social, and political referent bound up with childhood memories, group identities, and inter-generational and multi-ethnic tensions. Music is emotional both in terms of what happens to us when we listen to it and what certain musical styles stand for within a particular culture. Stefan Koelsch has outlined several reasons why music provided social benefits: (a) it provides a social contract between performer and listener and fellow listeners; (b) it offers the convergence of individual inferences as to the intentions and emotional states of others; (c) it enhances empathy and individual states become more homogeneous, which can decrease conflict, foster group identity, and confirm cultural models; (d) it increases the cognitive development of social communication; (e) it involves the coordination of actions and promotes synchronized movements; and (f) it can enhance cooperation, shared intentionality, and group cohesion.[66] If we accept this, then it should be equally accepted that when a musical form we associate with one or more of these benefits challenges our expectations, then, like the cognitive responses to narrative surprisal I outlined in Chapter One, we are likely to have an embodied emotional response. This can range from pleasurably surprising, to exciting, to absorbing and then to happiness, sadness, fear, and even anger.

There is also the biological effect of musical expectancy. This is shown by a recent small Swedish study on choral singing, which found that song structure and certain biological bodily functions were closely linked.[67] When tested, participants were found to have developed similar heart and respiratory rates depending on the structure of music they listened to, and that these rates tended to be lower than normal. These findings led the researchers to posit that singing in a choir had positive health benefits akin to the kind of breathing exercises practiced in yoga. This may be somewhat overstated, as this study only tested a small group of 15 healthy 18-year-olds and used three musical styles: a single tone, a hymn, and a chant. I wonder what the results would have been if the chorus members had been asked to sing a passionate lament? However, this study did help show that singing in a chorus is a collective biological experience as much as a social one, and the structure of the music has a profound effect on important human life mechanisms, including breathing and the heart.

Within Greek drama we also find the expectation that the right kind of music should fit the occasion. When it did not, we see tragic characters becoming deeply alarmed and adding to a sense of tension and anxiety. This "singing the wrong song" happens often in tragedy: in the *Agamemnon*, Cassandra is upbraided by a confused and worried chorus who tell the distraught prophetess to stop singing the lament to Apollo, who does not hear "cries of pain" (1150–1155). Later in the same scene she describes the Furies as sounding the celebratory *komos* as if they were supernatural guests at a drinking party at the House of Atreus, except she envisions them gorging themselves on blood. Finally, at the end of the trilogy in the *Eumenides*, Athena says she will never tire of making the Furies harmonize with her, both in terms of her judgment and music. Once they do "sing the same song" with Athena, they join a procession leading to the Acropolis and are welcomed to Athens as the "Sacred Goddesses" and the "Kindly Ones" (902–904).

Whereas we are currently deaf to the musical sounds of Greek drama, we can know something of the effect of the music by detecting moments of surprisal, dissonance, and disharmony within the texts of the plays by studying the metrical patterns of the language and the responses of the characters.[68] We can also understand a good deal of the effect of Greek dramatic music by knowing the kind of cognitive responses people generally have to particular instrumentation, rhythms, and pitch qualities. Though we cannot hear the music today, we can appreciate the enormous sensory and affective environment it provided in the Greek theatre and its profound contribution to the entire emotional experience. But in Greek drama, the music did not act alone; in the next chapter we will explore certain key performative aspects of words.

Notes

1 Pinker 1997: 534.
2 Plato, *Laws* 700b–701a; Aristotle, *Politics* 8.5–8; Pseudo Plutarch, 15; Plutarch, *Nicias* 29.3.
3 Pöhlmann, Egert, and West 2001; West 1992; Hagel 2009; Mathiesen 1999.
4 D'Angour, A., Song of the Sirens. www.armand-dangour.com/2014/03/song-sirens/. Accessed October 1, 2016.
5 Koelsch 2013: 286–303.
6 Higgins 2012: 273–282.

7 Fritz et al. 2009: 573–576.
8 Space precludes a full discussion on the controversial subject of basic emotions and the cross-cultural application of such theories. For a full discussion of the subject, see Plamper 2015, although he takes an overly simplistic view of the important work and impact of Ekman, Damasio, LeDoux, and others.
9 Zacharopoulou and Kyriakidou 2009: 1–15.
10 Zatorre and Salimpoor 2013: 10430–10437. For auditory dissonance in ancient Greece, see Gurd 2016.
11 Juslin 2011: 113–135. See also Simpson et al. 2008: 596–597.
12 Zatorre and Salimpoor 2013: 10430–10437; Conard et al. 2009: 737–740.
13 Münzel et al. 2002: 107–110.
14 D'Angour has compared the sound of the aulos to that of the Sardinian Launeddas. D'Angour 2017: 428–444. To hear the Launeddas, visit www.youtube.com/watch?v=BzWITU3L-F0. Accessed June 18, 2016. For "buzzing" Aristophanes, *Acharnians* 864–866; for "geese-screeching" Hsch. s.v. *krizei*.
15 See Declercq and Dekeyser 2007: 2018–2019 and Fametani et al. 2008: 1557–1567.
16 Plato, *Symposium* 215c; *Minos* 318b.
17 Aristotle *Politics* 1339.
18 Wilson 1999: 58; Barker 1989: 15–16. I am using the term *aulos* to describe the twin reed pipe instrument that is also termed *auloi* by some scholars. On the aulos in general, see Hagel 2009: 327–343; West 1992: 81–107; Mathiesen 1999: 177–222.
19 Pausanias 9.12.5.
20 Plutarch *Life of Alcibiades* 2.4–6.
21 Wilson 1999: 87–92.
22 Plato, *Protagoras* 347(d); Sophocles, *Trachiniae* 640–641; Euripides, *Bacchae* 127–128.
23 Hagel 2009: 328–329.
24 Aristoxenus says there are three octaves between *parthenoi* and *hyperteleioi* (fr. 101 & *El. Harm.* 20.32–34). Citing Aristoxenus, Athenaeus says that the *teloi* and *hyperteloi* are men's auloi (*Deipnosophistae* 4.79).
25 Pindar *Pythian* 12, *Olympian* 12., *Isthmian* 5.27.
26 Simpson et al. 2008: 596–595.
27 Panksepp 1995: 171–207; Rickard 2004: 371–388.
28 Lysimachus ruled from 306–281, so this was a distant event for Lucian, who was writing in the second century CE.
29 Wilson 2005: 183–193.
30 See Wright 1986; Suter 2008: 156–180.
31 Many references to lamentation in Greek drama have been collected in Barker 1989: 63–71.
32 Holst-Warhaft 2002: 98–126.
33 Murray 1913.
34 Nilsson 1951: 61–145.
35 Else 1972.
36 Pickard-Cambridge et al. 1968: 9–10.
37 Alexiou 2002. See also Dué 2006; Foley 2009.
38 Loraux 1998: 7.
39 Plutarch, *Life of Nikias*, 29, 2–3.
40 Hume 1907: 216–217. See Dadlez 2004: 213–236.
41 Aristotle also makes this point in *Politics* 8.1342a 1–10.
42 Cairns has explored the Greek concept of *phrike* in tragedy. Cairns 2015: 75–94.
43 Greenberg et al. 2015. Also see Chapter Seven.
44 Juslin 2013; Juslin et al. 2014.
45 Ondobaka et al. 2015: 2.
46 Seth 2013:565–573.

47 Prinz 2012: 242–247.
48 Ondobaka et al. 2015: 4.
49 Meineck 2012: 7–24. Prolactin can also be activated by music. Levitin and Tirovolas 2009 have shown that Prolactin is a tranquilizing hormone released by the anterior pituitary gland when a person feels sad and that can be activated when people listen to sad music. Huron 2011: 146–158. Plato wrote that the soul "hungers for the satisfaction of weeping and wailing" and this "is the very part that receives satisfaction and enjoyment from poets" (*Republic*. 10.606a). See also Levitin 2013.
50 Loraux 1998: 7. See also, Berthomé and Houseman 2010: 57–75.
51 Schubert 1996.
52 Seligman and Kirmayer 2008.
53 Garrido and Schubert 2011.
54 Butler 2006.
55 Agarwal and Karahanna 2000.
56 Athenaeus, *Deipnosophista*i 628a–b.
57 Seligman and Kirmayer 2008.
58 Seligman and Kirmayer 2008: 35.
59 Gebauer et al. 2015: 50–52.
60 Dionysius of Halicarnassus wrote that Euripidean music was more intricate and elaborate than the text alone indicates and that the music had become independent from the tonal pitch qualities of Greek speech. *On Literary Composition* (11.42UR).
61 Translated by Trevor J. Saunders. *Laws* 812d: "they must produce notes that are identical in pitch to the words being sung." The Lyre should not be used to play an elaborate independent melody.
62 D'Angour 2006: 264–283.
63 On the "New Music" in Greek drama, see Csapo 2004: 207–248; Franklin forthcoming; Leven 2010: 35–48; D'Angour 2006: 264–283.
64 Csapo 2004: 207–248.
65 Aristophanes, *Thesmo.* 100, *Suda*. SV Philoxenus; Pseudo Plutarch *On Music* (1142a).
66 Koelsch 2015: 193–201.
67 Vickhoff et al. 2012.
68 For example, Scott 1984 and 1996; and Goldhill 2012: 13–136.

Bibliography

Agarwal, R. and Karahanna, E. 2000. "Time flies when you're having fun: Cognitive absorption and beliefs about information technology usage" in *Management Information Systems Quarterly* 24.4: 665–694.

Alexiou, M. 2002. *The ritual lament in Greek tradition*. Lanham, MD: Rowman & Littlefield.

Barker, A. 1989a. *Greek musical writings, vol. 1: The musician and his art.* Cambridge: Cambridge University Press.

Barker, A. 1989b. *Greek musical writings, vol. 2: Harmonic and acoustic theory.* Cambridge: Cambridge University Press.

Berthomé, F. and Houseman, M. 2010. "Ritual and emotions: Moving relations, patterned effusions" in *Religion and Society: Advances in Research* 1.1: 57–75.

Butler, L.D. 2006. "Normative dissociation" in *Psychiatric Clinics of North America* 29: 45–62.

Cairns, D. 2015. "The horror and the pity: Phrikē as a tragic emotion" in *Psychoanalytic Inquiry* 35.1, 2: 75–94.

Conard, N.J., Malina, M. and Münzel, S.C. 2009. "New flutes document the earliest musical tradition in southwestern Germany" in *Nature* 460.7256: 737–740.

Csapo, E. 2004. "The politics of the new music" in Murray, P. and Wilson, P. (Eds.), *Music and the muses: The culture of 'mousikē' in the classical Athenian city*. Oxford: Oxford University Press: 207–248.

Dadlez, E.M. 2004. "Pleased and afflicted: Hume on the paradox of tragic pleasure" in *Hume Studies* 30.2: 213–236.

D'Angour, A. 2006. "The new music – so what's new?" in Goldhill, S. and Osborne, R. (Eds.), *Rethinking revolutions through ancient Greece*. Cambridge: Cambridge University Press: 264–283.

D'Angour, A. 2017. "Euripides and the sound of music" in McClure, L. (Ed.), *The Blackwell companion to euripides*. New York: Wiley-Blackwell: 428–444.

Declercq, N.F. and Dekeyser, C.S.A. 2007. "Acoustic diffraction effects at the Hellenistic amphitheater of Epidaurus: Seat rows responsible for the marvelous acoustics" in *Journal of the Acoustical Society of America* 121.4: 2018–2019.

Dué, C. 2006. *The captive woman's lament in Greek tragedy*. Austin, TX: University of Texas Press.

Else, G.F. 1972, *The origin and early form of Greek tragedy*. New York/London: Norton.

Fametani, A., Prodi, N.P. and Pompoli, R. 2008. "On the acoustics of Greek and Roman theaters" in *Journal of the Acoustical Society of America* 124.3: 1557–1567.

Foley, H.P. 2009. *Female acts in Greek tragedy*. Princeton, NJ: Princeton University Press.

Franklin, J.C. Forthcoming. "Dithyramb and the demise of music" in Kowalzig, B. and Wilson, P. (Eds.), *Song culture and social change: The contexts of dithyramb*.

Fritz, T., Jentschke, S., Gosselin, N., Sammler, D., Peretz, I., Turner, R., Friederici, A.D. and Koelsch, S. 2009. "Universal recognition of three basic emotions in music" in *Current Biology* 19.7: 573–576.

Garrido, S. and Schubert, E. 2011. "Individual differences in the enjoyment of negative emotion in music: A literature review and experiment" in *Music Perception: An Interdisciplinary Journal* 28.3: 279–296.

Gebauer, L., Kringelbach, M.L. and Vuust, P. 2015. "Predictive coding links perception, action, and learning to emotions in music: Comment on 'The quartet theory of human emotions: An integrative and neurofunctional model' by S. Koelsch et al." in *Physics of Life Reviews* 13: 50–52.

Goldhill, S. 2012. *Sophocles and the language of tragedy*. Oxford/New York: Oxford University Press.

Greenberg, D.M., Rentfrow, P.J. and Baron-Cohen, S. 2015. "Can music increase empathy? Interpreting musical experience through the empathizing – systemizing (ES) theory: Implications for Autism" in *Empirical Musicology Review* 10.1–2: 80–95.

Gurd, S.A. 2016. *Dissonance: Auditory aesthetics in ancient Greece*. New York: Fordham University Press.

Hagel, S. 2009. *Ancient Greek music: A new technical history*. Cambridge: Cambridge University Press.

Higgins, K.M. 2012. "Biology and culture in musical emotions" in *Emotion Review* 4.3: 273–282.

Holst-Warhaft, G. 2002. *Dangerous voices: Women's laments and Greek literature*. London: Routledge.

Hume, D. 1907. *Essays: Moral, political, and literary* (Vol. 1). London: Longmans, Green, and Company: 216–217.

Huron, D. 2011. "Why is sad music pleasurable? A possible role for prolactin" in *Musicae Scientiae* 15.2: 146–158.

Juslin, P.N. 2011. "Music and emotion: Seven questions, seven answers" in Deliège, I. and Davidson, J. (Eds.), *Music and the mind: Essays in honour of John Sloboda*. Oxford: Oxford University Press: 113–135.

Juslin, P.N. 2013. "From everyday emotions to aesthetic emotions: Towards a unified theory of musical emotions" in *Physics of Life Reviews* 10.3: 235–266.

Juslin, P.N., Harmat, L. and Eerola, T. 2014. "What makes music emotionally significant? Exploring the underlying mechanisms" in *Psychology of Music* 42.4: 599–623.

Koelsch, S. 2013. "Emotion and Music" in Armony, J. and Vuilleumier, P. (Eds.), *The Cambridge handbook of human affective neuroscience*. Cambridge: Cambridge University Press: 286–303.

Koelsch, S. 2015. "Music-evoked emotions: Principles, brain correlates, and implications for therapy" in *Annals of the New York Academy of Sciences*1337.1: 193–201.

Leven, P.A. 2010. "New music and its myths: Athenaeus' reading of the Aulos revolution (*Deipnosophistae* 14.616 e – 617f)" in *The Journal of Hellenic Studies*130: 35–48.

Levitin, D.J. 2013. "Neural correlates of musical behaviors: A brief overview" in *Music Therapy Perspectives* 31.1:15–24.

Levitin, D.J. and Tirovolas, A.K. 2009. "Current advances in the cognitive neuroscience of music" in *Annals of the New York Academy of Sciences* 1156.1: 211–231.

Loraux, N. 1998. *Mothers in mourning*, C. Pache (Tr.). Ithaca, NY: Cornell University Press.

Mathiesen, T.J. 1999. *Apollo's lyre: Greek music and music theory in antiquity and the middle ages* (Vol. 2). Lincoln, NE: University of Nebraska Press.

Meineck, P. 2012. "Combat Trauma and the tragic stage: 'Restoration' by Cultural Catharsis" in *Intertexts* 16.1: 7–24.

Münzel, S.C., Seeberger, F. and Hein, W. 2002. "The Geißenklösterle Flute – Discovery, experiments, reconstruction" in Hickmann, E., Kilmer, A.D. and Eichmann, R. (Eds.), *Studien zur Musikarchäologie III: Archäologie früher Klangerzeugung und Tonordnung; Musikarchäologie in der Ägäis und Anatolien, Orient-Archäologie*, Bd. Rahden: Verlag Marie Leidorf GmbH: 107–110.

Murray, G. 1913. *Euripides and his age*. London: Henry Holt.

Nilsson, M.P. 1951. *Cults, myths, oracles, and politics in ancient Greece* (Vol. 1). Lund, Sweden: CWK Gleerup.

Ondobaka, S., Kilner, J. and Friston, K. 2017. "The role of interoceptive inference in theory of mind" in *Brain and Cognition* 112: 64–68.

Panksepp, J. 1995. "The emotional source of 'chills' induced by music" in *Music Percept* 13.2: 171–207.

Pickard-Cambridge, Sir A.W., Gould, J.P. and Lewis, D.M. 1968. *The dramatic festivals of Athens . . . Revised by John Gould and DM Lewis*. Oxford: Clarendon Press.

Pinker, S. 1997. *How the mind works*. New York: Norton.

Plamper, J. 2015. *The history of emotions: An introduction*. K. Tribe (Tr.). Oxford: Oxford University Press.

Pöhlmann, E. and West, M.L. (Eds.). 2001. *Documents of ancient Greek music: The extant melodies and fragments*. Oxford: Oxford University Press.

Prinz, J.J. 2012. *Beyond human nature: How culture and experience shape our lives*. London: Penguin: 242–247.

Rickard, N.S. 2004. "Intense emotional responses to music: A test of the physiological arousal hypothesis" in *Psychol Music* 32.4: 371–388.

Schubert, E. 1996. "Enjoyment of negative emotions in music: An associative network explanation" in *Psychology of Music* 24: 18–28.

Scott, W.C. 1984. *Musical design in Aeschylean theater*. Lebanon, NH: Dartmouth College Press.

Scott, W.C. 1996. *Musical design in Sophoclean theater*. Lebanon, NH: Dartmouth College Press.

Seligman, R. and Kirmayer, L. 2008. "Dissociative experience and cultural neuroscience: Narrative, metaphor and mechanism" in *Culture, Medicine and Psychiatry* 32.1: 31–64.

Seth, A.K. 2013. "Interoceptive inference, emotion, and the embodied self" in *Trends in Cognitive Sciences* 17.11: 565–573.

Simpson, E.A., Oliver, W.T. and Fragaszy, D. 2008. "Super-expressive voices: Music to my ears?" in *Behavioral and Brain Sciences* 31.5: 596–597.

Suter, A. 2008. "Male lament in Greek tragedy", in Suter, A. (Ed.), *Lament: Studies in the ancient Mediterranean and beyond*. Oxford: Oxford University Press: 156–180.

Vickhoff, B., Malmgren, H., Aström, R., Nyberg, G., Ekström, S.R., Engwall, M., Snygg, J., Nilsson, M. and Jörnsten, R. 2012. "Music structure determines heart rate variability of singers" *Frontiers in Psychology* 4.334: 1–16.

West, M.L. 1992. *Ancient Greek music*. Oxford: Oxford University Press.

Wilson, P. 1999. "The aulos in Athens" in Goldhill, S. and Osbourne, R. (Eds.), *Performance culture and Athenian democracy*. Oxford: Oxford University Press: 58–95.

Wilson, P. 2005. "Music" in Gregory, J. (Ed.), *A companion to Greek tragedy*. Malden, MA: Blackwell: 183–193.

Wright, E.S. 1986. *The form of laments in Greek tragedy*, Doctoral Dissertation, University of Pennsylvania.

Zacharopoulou, K. and Kyriakidou, A. 2009. "A cross-cultural comparative study of the role of musical structural features in the perception of emotion in Greek traditional music" in *Journal of Interdisciplinary Music Studies* 3.1: 1–15.

Zatorre, R.J. and Salimpoor, V.N. 2013. "From perception to pleasure: Music and its neural substrates" *Proceedings of the National Academy of Sciences* 110 (Supp. 2): 10430–10437.

6 *Lexis*

Somatosensory words

In the *Birth of Tragedy*, Nietzsche wrote that Greek tragedy presents itself to us "only as word-drama" and made a powerful case for how the inherent spirituality of tragedy was conveyed by the music *with* words; hence his original publication included in its title, *From the Spirit of Music*.[1] Nietzsche was quite right – in terms of the experience of witnessing the performance of a Greek drama, its words cannot be separated from its music, and yet we do not have the music, only the words, and most people encounter the art form through reading, not listening. The purpose of this study is not to denigrate the enormous contributions that philology and literary theory have made to our knowledge and appreciation of ancient drama, but from a performative perspective it is also important to explore the experiential aspects of the drama that also contributed to the *Gesamtkunstwerk* that was classical Athenian theatre. These plays were originally created to be uttered live in action, either recited as formal metered poetry, usually iambics, or sung as lyrics set to music. As Hall has pointed out, "in antiquity . . . the relationship of the art of acting to the art of singing was often inextricable."[2]

The way in which language was used in the theatre was as important for Aristotle as it is for theatre practitioners today. He rightly connects language with *dianoia* (thought) as an effective instrument for advancing the intentions and objectives of the plot (*Poetics* 1450b12–14). A great deal has and will be written on the meaning of the words of Greek drama, but my intention here is to examine *lexis* from a performative perspective. The words of Greek drama as transmitted by the textual tradition have been the way in which we have come to know these plays, and there remains an important and continued need to analyze and understand the narrative action of these works and the political, social, and cultural meanings of their words. Here I am aiming to examine some of the affective qualities of verbal utterances beyond their function as a linguistic representational system. These are the cognitive properties of meter, the effect of perceiving speech and song production from a mask, the effect of linguistic evaluative conditioning, and the affective power of unintelligible utterances.

There is now a good deal of evidence to suggest that lyrics enhance music's ability to evoke certain emotions. This is supported by a recent fMRI study that suggested that sad music is perceived as far sadder with lyrics than without.[3] Yet the same is not true for happy music. In this study the participants found that the music seemed happier without lyrics than with. Perhaps this points to the elements of self-identification, projection, and empathy that lyrics can help contextualize and personalize. Aristotle takes up this same theme in *Problemata* 19.6 (918a10–13): "Why does recitation with a musical accompaniment have a tragic effect when introduced into singing? Is it owing to the resulting contrast? For the contrast gives an expression of feeling and implies extremity of calamity or grief, whereas uniformity is less mournful." Surely, what Aristotle is referring to here is the combined cognitive effect of words delivered with music. Though we do not have the music, the words we have can provide some evidence of the emotional experiences that words with music offered.

The movement of meter

Nietzsche was right to point out that the words of ancient drama did not operate in isolation from the multisensory environment of the Festival of Dionysos. By "multisensory" I mean everything available to the cognitive system of each original audience member that was synthesized into a holistic experience that unfolded as the performance happened. The close reading of a play is a totally different cognitive experience than the performance of the play, where words are spoken and heard in the moment. What cannot be discerned by reading alone is that in performance the predominantly higher-order cognitive processing of the spoken word is subordinated to the lower-order processing of visual or other somatosensory data.

A demonstration of this in action is the McGurk effect, named after developmental psychologist Harry McGurk.[4] One person is asked to mime "Ba, Ba, Ba" over and over again while another person standing behind him and, out of view, speaks the words "Ba, Ba, Ba" at the same time. Then the miming person is told to change her mime and mouth "Fa, Fa, Fa" again and again while the speaking person continues saying "Ba, Ba, Ba." The audience then hears "Fa, Fa, Fa" even though what the speaking person is actually saying is "Ba, Ba, Ba." The audience's visual processing system trumps the audience's auditory processing system.[5]

The McGurk effect is important because it reminds us that spoken or sung text is not the primary cognitive element in a live theatre production; its meaning can be usurped by the visual stream, and the text alone cannot reveal the incongruities and dissonance that may have resulted from the accompanying music or visual action that may have fundamentally altered the meaning of the words in performance. As Philip Auslander has pointed out, the text is a blueprint for performance – it is but one, albeit an important one, part of the realization of a performance. This is why the two best words a director ever hears in a rehearsal room is when the actor declares she is "off book" – the written text has become speech and song embodied and enlivened by the movements and gestures of the performer.

Poetic speech and song can add another level of affective communication beyond the meaning of the words being uttered. The sound of language and how it is organized in its delivery is a product of bodily movement, hence verbal utterances can be described as auditory movements in that they rely on the movement of the vocal cords and mouth and air through the larynx to generate sound. Music and language also share the same auditory pathways, whereby the sonic input travels from the cochlea to the cortex via the brainstem and the inferior colliculus and share executive function motor pathways, which are involved in body movements. It has also been shown that areas of the brain that have been associated with language processing, such as Wernicke's and Broca's areas, are similarly activated by music.[6]

Aristotle wrote, "Sound is the movement of that which can be moved" (*On the Soul* II, 8, 420a19), and neuroscientist and musicologist Robert Zatorre has made the same observation, pointing out the fact that "music is movement."[7] What he means is that the motor systems of the human brain, particularly those that control fine movements, are required for the musician to produce the delicate physical movements needed to make music. Also, output signals emanating from the pre-motor cortex affect responses within the brain's auditory cortex. A complex neural network is at work when we both play and listen to music, one that is linked to our brain's pre-motor and motor systems. The vibration of a string, the pulsation of a drumbeat, the forcing of air through a pipe – music *is* movement enacted and music makes us move.

The words of Greek drama were set to distinctive metrical patterns; Sarah Nooter has commented how the mixture of meters in tragedy is "one of the genre's most interesting and innovative qualities," and Simon Goldhill has described meter as "part of the signifying repertoire of tragedy."[8] The meter of poetic speech and song forms a grammar of auditory movement in how the utterances are arranged to be delivered. This can be equated to what Juslin has described as rhythm entrainment.[9] This occurs when the external tempo of the music influences the internal rhythms of the body, such as heart and respiratory rates, evoking proprioceptive feedback and states of arousal. The slower tempos and discordant tonal qualities of sad music thus affect our physical state in terms of heart rate, respiration, and how we might move, or think about moving, to the music. Zatorre et al. have pointed out that "psychophysiological changes that are associated with listening to music might also be a byproduct of the engagement of the motor system, and therefore would also provide afferent [inward carrying] feedback enhancing the affective state."[10] Entrainment has also been shown to promote social bonding in that humans tend to synchronize some of their biological activity, such as respiratory and heart rates, to external rhythms. When these rhythms are experienced collectively, humans are better disposed to feelings of group solidarity.[11] We have already explored kinesthetic empathy and neural responses to gestures and dance; when connected with rhythm, either of the music, the meter of the lyrics, or both, these can be very powerful cognitive transmitters of emotion, particularly amongst the kind of collective gathering that is a theatre audience.

Brain imaging studies show us that the listener processes the sounds of words, or their "musicality," in the primary auditory cortex, the area of the brain that responds to instrumental music. This region is not used for processing words and lyrics. Furthermore, this kind of auditory imagery seems to use areas of the brain responsible for motion that are partially located in the cerebellum. The cerebellum is also involved in temporal coordination of both external movements (e.g., being able to adequately make timing predictions about when to move to catch a falling object) and internalized "movements" produced by the metrical patterns of poetic speech.

The cognitive effect of meter was well understood in antiquity. The Sicilian rhetorician Gorgias wrote a treatise called *The Encomium of Helen*, which demonstrated the power of persuasive speech. In it he argued that Helen could not be blamed for the Trojan War, as she was enthralled by "speech." Gorgias makes it clear that the most affective form of persuasive speech is poetry and names it "discourse having meter," adding that "fearful shuddering and tearful pity and sorrowful longing come upon those who hear it, and the soul experiences a peculiar feeling, on account of the words, at the good and bad fortunes of other people's affairs and bodies."[12] Here, Gorgias describes something similar to Aristotle's theory of *catharsis* and the emotional responses to drama provoked by emotions such as pity and fear. Like Gorgias, Aristotle also equates these affective concepts to poetic meter in *Poetics* (1449b, 28–30).

Gorgias's vivid embodied description of the emotional effects of poetry is an ancient concept of mimetic empathy and very close to modern theories of embodied cognition and the empathetic responses of the human sensorimotor system. On the other hand, Plato relegated meter to the sidelines, describing it as a mere "musical coloring" (*Rep.* 601a-b). Coincidently, in his dialogue, *Gorgias*, he complains that poetry is just rhetoric dressed up in meter and music. Plato maintained, that although poetry feels good it may well not be food for you. He compared drama and music to the confections of a pastry chef as they appeal to the appetite. Thus, music and drama must be curtailed, just as a doctor advises that the patient refrain from eating pastries in order to heal (502c).

This brings us back to Steven Pinker's famously dismissive pronouncement that music is just "auditory cheesecake" and has no evolutionary adaptive value.[13] Pinker compares music to "prosody" – the cadence, rhythm, stresses, and intonation of human speech – the very elements that poetic speech emphasizes. He surmises that music (and by the same argument, poetry) has no real adaptive purpose in our survival but mimics several other evolved adaptive behaviors such as language, auditory scene analysis, emotional calls, habitat selection, and motor control. In other words, it is just like the confections of Plato's pastry chef and serves no other purpose than giving pleasure.

Plato and Pinker may have both been right about the effect of poetry and music but totally wrong about its importance in human culture and cognitive processes. For example, a recent German study examined people responding to four different stanzas of poetry.[14] The first was both metered and rhymed, the second just metered, the third non-metered but rhyming, and the fourth non-metered and nonrhyming. As Greek did not use rhymes, we will focus on the two non-rhyming stanzas.

Metered

Holdes Bittern, mildes Begehren,
wie es süß dem Herzen klingt!
Durch die nacht, die mich umfangen,
Blickt zu mir der Töne Licht!

Non-metered

Holdes Bittern, mildes Begehren,
wie es süß mir im Herzen klingt!
Durch die Dunkelheit, die mich umfangen,
Blickt zu mir deiser Töne Licht!

It actually helps if you don't know German well and just focus on the meter of the two poems. For most people the first is easier to read (and hear), whereas the second takes a while longer to process. The study showed that metered poetry led to a reported enhanced aesthetic appreciation, and emotionality was reported to be greater in the metered than non-metered poems. This may be due to the meter-increased prediction fluency – the ability to quickly and more effortlessly process complex information. The pleasure of poetry as a kind of verbalized music" to read actually has a serious cognitive purpose.

A recent brain imaging study using positron emission tomography (PET) and fMRI scanning showed that even though music is an abstract stimulus, it can create euphoric feelings and cravings identical to real rewards involving the brain's striatal dopaminergic system.[15] This study indicated an increase in levels of dopamine in the striatum of the listener at moments of peak emotional arousal experienced in the music. These moments were corroborated, with the participants reporting moments of "chills" or frissons. This new research is a verification of earlier theories of musical expectancy, which have been connected to our predictive process, and it goes some way toward explaining the physiological effects of the pleasures of music and poetry.

Why then is the brain signaling our reward centers when we hear music or poetry? While Pinker's famous statement has been widely attacked, and Plato's attitudes toward music and poetry are regarded as artistic conservatism, the fact that music and poetry so profoundly affect the neural chemistry of the brain suggests that they have evolved to efficiently activate the emotional processing parts of the human brain and body and to rapidly communicate commonly shared cultural attitudes. This is something very much like Gorgias's idea that "the soul experiences a peculiar feeling, on account of the words, at the good and bad fortunes of other people's affairs and bodies." Vuust and Kringelbach have shown how music and language share common neural processing features and work by way of predictive mechanisms in that they establish, confirm, or disrupt anticipatory structures. They state that "the hierarchical structure of the meter underlies all other expectancy structures in music such as rhythm, harmony, melody, and intensity, in that it influences perception of any musical event. Hence, anticipatory structures such as the meter (but also, for example, tonality) provide the listener with a framework for interpreting and remembering music."[16] They surmise that it is impossible to memorize a complex rhythm if you do not know the meter.

In *Poetics* Aristotle also connects words and music when he describes mimetic art as made up of *ruthmos* (rhythm), *logos* (speech), and *harmonia* (music or tune), alone or in any combination (1447a23). In the ancient Greek theatre, poetic speech and music were two major factors in drama's ability to move the human "soul" – and while we have lost nearly all trace of the music, the poetry and meter of the ancient Greek is still with us and can be discerned from the surviving texts, affording us the opportunity to understand at least the narrative and rhythmic means by which the music and language were conveyed. Vuust and Kringelbach highlight how music, language, reading, and writing share fundamental neural mechanisms and "allow us to communicate, record, experience and imagine the hedonic experiences of other humans across time and space."[17] Perhaps this helps frame Aristotle's comment that the vividness (*enargês*) of tragedy, which provides so much pleasure, is enhanced by *opsis* and *mousikê* but could be felt whether the play was watched or read (*Poetics* 1462a-b). Thus, as long as the study of ancient Greek remains an important part of our own academic cultures, then we can continue to access this fundamental experiential aspect of ancient drama.

Masked language

One of the main objectives of this study is to apply cognitive methodologies to place ancient drama in its performative context. If we follow this same approach to the way in which the lyrics were received, then we should consider singing or speaking these words in a mask. One of the most important processes that occurs when we process speech is the visual information conveyed by the anatomical working of speech (the movement of the mouth, lips, tongue, jaw, and throat) and the mental mapping of faces, even when listening to a disembodied voice. Have you ever visualized a radio presenter and then been quite shocked when you finally see an image of the person? If this is the case, then what happens when the voice you are processing emanates from a mask? Is language enhanced or diminished, and how was it processed by audience members hearing it live in a large open-air theatre, without amplification, by actors wearing full-face masks? How did the poetic and musical cadences of Greek dramatic language affect this experience?

First, let us consider how we process speech and attend to the speaker. This has been studied at least since the 1950s, when research in this area focused on air traffic controllers who had to discern the speech of the pilot from the plane they were tracking from among a cacophony of other pilots over a radio system and the vocal noise of other speaking controllers in the room.[18] This came to be called "the cocktail party effect." This term describes how humans are able to focus their auditory attention on a particular speaker, even though many other spoken conversations are going on all around them, such as at a cocktail party. Other research at that time examined the related phenomena of being able to pick out certain key words from "background conversations" such as one's own name or taboo and offensive terms.[19] This human auditory ability, which has been called "selective attention"[20] or the "attenuation model,"[21] has been recently summed up in a study by Nima Mesgarani and Edward Chang:

> Concurrent complex sounds, which are completely mixed upon entering the ear, are re-segregated and selected from within the auditory system. The resulting percept is that we selectively attend to the desired speaker while tuning out the others.[22]

This particular study indicated that the cortical representation of speech does not just mirror the external acoustic environment, but instead restores the representation of the speaker being attended to and suppresses the competing words, which are considered irrelevant. This is an innately human ability, and to date machine-based speech processors cannot replicate it.

What we seem to be actually doing here is mapping the speech that we hear onto a "cognitive face" that we create in order to better process speech that is difficult to understand or muffled by other external or distracting noises. Speech perception is thus a multisensory process involving auditory, visual, and haptic cognitive channels – each one compensating for the other when speech is compromised and each one contributing to how we build up a cognitive "picture" of the speech that we hear. When we view the person speaking to us, his or her facial kinematics play an important part in speech comprehension.[23] The perceived movement of the speaker's lips, mouth, tongue, jaw-line, throat, and other facial features have all been shown to vastly improve speech communication. Notice that when we want to mime something important to another person when we need to be quiet, such as at the theatre to a friend in the next row, we accentuate our facial features so the recipient can lip-read what we are saying. Actors performing in large theatres without microphones also over-articulate consonants and enlarge their facial speech movements while doing so, in order to communicate over large distances. But in the Greek theatre, the actors' faces were not seen and their mouths were hidden behind a mask, so did the mask hinder the comprehension of the words of Greek drama or perhaps enhance it?

A study published in 2014 by Sonja Schall and Katharina von Kriegstein proposed that the brain uses a simulation of the speakers talking face to optimize auditory perception.[24] They used fMRI to show that face movement sensitive areas in the left posterior superior temporal sulcus communicate with auditory speech processing sensitive areas in the left anterior superior temporal gyrus and sulcus when participants listened to auditory-only speech. From studies such as this it seems as if the brain does use a simulation of a face to optimize auditory perception even when listening to disembodied speech. This study also showed that when the speaker's voice was known to the participants, speech comprehension increased. Based on these studies and other cognitive research, it seems that wearing a mask might have been an audible disaster – but perhaps not.

Another study from Australia used mask-like computer-generated faces to come to the same conclusion about the combination of recognizable voice and recognizable face leading to the best speech perception.[25] This begs the question: how far did the ancient audience identify the voice of the actor with the face of the mask? Was it even important to be able to cognitively map the human face of the actor while watching the mask he was wearing? In my own experience working with masks in performance, both open-air and in interior spaces, this mask/voice issue has been a problem. Audiences have difficulty identifying which mask is speaking and can find it hard to "tune-in" to the words emanating from it. The mask, in effect, seems to short-circuit the normal speech processing channels. Would this problem have been negated if the audience knew the actor behind the mask and could mentally perceive his face speaking? We do know that the actors of Athenian drama were required to appear before an audience prior to the performance at an event called the *proagon*. Here the actors were introduced to an audience and the themes of the play were outlined.[26] Perhaps this pre-play event served to help the subsequent processing of masked speech in the actual performance.

Oliver Taplin has commented that dramatic speech is actually enhanced by the mask, and when he watches masked performances he pays attention to the words and their meanings more closely.[27] Taplin cites an essay by British theatre director Peter Hall, who had worked with masks made by Jocelyn Herbert in several productions, most memorably his *Oresteia* at the National Theatre in 1981. Hall compares the mask to the iambic pentameter of Shakespeare or the "formal grammar of Mozart's music" in that it creates a theatrical discipline, a constraining framework within which the play operates, which actually strengthens the expressive qualities of the performance, or as Hall puts it, "it releases rather than hides: it enables emotion to be specific rather than generalized."[28] Hall attributes a strength to the ambiguity of speech in a mask, especially in terms of the chorus. For him, one spoken voice is shared by the reactions of the entire group, and the fact that we may not be able to identify the speaker is a strength. This is an interesting theory, but how might that apply to the actors engaged in dialogue in two- and then three-handed scenes with choral interjections?

In his *Oresteia*, Hall used the highly structured Anglo-Saxon cadences of Tony Harrsion's poetic speech underscored by a stark percussive score by Harrison Birtwhistle. Poetic speech and song benefit the mask, their meters and over-articulated phonetic gestures help project the words. In this sense the mask may actually provide a stronger multi-modal cognitive platform for song and poetic speech: the same cognitive processes that allow the individual spectator to project emotional states onto the ambiguous surface of the mask's features may do the same with language. The mask may have enhanced and personalized speech perception by "telescoping" language processing by means of highly coordinated movement, gestures, and heightened poetic language. Future studies on facial mapping and speech processing might consider comparing neural responses to speech generated by a human face to speech coming from a mask. My hypothesis is that the visual facial mapping areas of the brain shown to be active in speech perception would be far more active when watching the mask and the experience more intense and far more personal.

When speaking through a mask, movement is key, and this must have also been true of the ancient mask. In 2011 Desiree Sanchez directed a production of Pirandello's *Six Characters in Search of an Author* using masks made by David Knezz and constructed based on my research. This metaphysical play, first performed in Italy in 1921, examines the nature of theatrical reality and the propriety of narrative, when six members of a phantom family invade a rehearsal room and demand that the theatre company there perform their story. In the first restaging of the play, Pirandello placed his six characters in masks playing against the "actors" in the play who performed bare-faced. He called his masked characters "immutable creatures of the imagination, and therefore more consistent and real than the ever-changing naturalness of the actors".[29] This is similar to Hall's notion that the mask provides a constraining frame, which enhances audience perception.

In working with the masks in rehearsal, Sanchez found that the words the masked actors spoke needed to be framed by physical gestures. Audience attention needed to be directed toward the masked speaker by both a framing movement performed by the actor and an attentional (but not necessarily directional) focus on the part of the other performers on stage. Speech also needed to be heightened, over-articulated, and very precise, and every micro-movement of the actor needed to be carefully coordinated. Any lack of physical continuity with the words would be quickly detected, and this visual/aural discontinuity would seriously hinder audience perception of the words. Look at the images of masked actors on the surviving vase painting again; in performance their movements of hands and feet are precise and over-articulated in a kind of gestural ballet (Chapter Four, Figures 4.1 and 4.2).

Such theatre practices are supported by a recent Chinese study, which summed up speech perception thus:

> Speech perception is not just for hearing speech sounds, but more essentially, for recognizing and understanding speech signals, requiring that multisensory modalities interact. In fact, speech understanding and speech hearing do not share the same brain network, including the motor areas.[30]

Thus we can conclude that acting in a mask may well have accentuated the reception of the words of drama by enhancing the multi-modal and sensorimotor facets of speech processing and movement.

Linguistic evaluative conditioning

In addition to the multisensory experiences that were present and active in the audiences' sensory fields, words in the form of lyrics and poetic language can evoke another subjective level of sensation.[31] Language that describes or sounds like a particular sensory experience activates the same areas of the brain associated with tangible sensory inputs and the processing of working memory. Juslin has called this process evaluative conditioning and suggests that it occurs when a particular piece of music is associated with a distinct emotional state. If this pairing of melody and specific emotion is repeated enough in the music and meter, then the action of hearing that refrain will start to evoke the same emotional response. One well-known example of evaluative conditioning is Wagner's purposeful use of the leitmotif in opera. Though the melodies of Greek theatrical music are lost to us, we know that they tended to be performed in certain well-known musical modes. For example, in *The Republic*, Plato describes lamentation as usually accompanied by the mixed Lydian or what he explains as the "tense and shrill" Lydian mode (3. 398e).[32] The lyrics of drama also involve repeated phrases or refrains that act in a similar cognitive fashion as Wagnerian leitmotifs. For example, the chorus of Aeschylus's *Agamemnon* sings the same evocative refrain three times in their first choral song (121,138 and 159):

Ailinon ailinon eipe, to d' eu nikatō.
Cry, cry the song of sorrow, but let the good prevail.

"*Alinon, alinon*" was a commonly known lyrical motif in Greek song and was called the "lament for Linus." He was the brother of the mythical Orpheus, also a musician, and in the mythic tradition is killed by Herakles. The song is found in Homer and is depicted being performed at a harvest celebration on the shield of Achilles in *The Iliad* (18.570–572). This repeated phrase was a recognizable cry of lamentation for the Athenian audience and can also be found in Sophocles' *Ajax* (624–635) and Euripides' *Orestes* (1395). The refrain of *alinon, alinon* in the Linus song operates on several multisensory levels: the sound of the word with its long vowels mimics an unintelligible cry of mourning, the recognition of the name of Linus anchors this mimetic moment to a well-known tragic story, and the repetition of this lyrical motif primes the evaluative conditioning of the audience.

Tragic choral song in particular is replete with structural leitmotifs supported by the antiphonal strophe/antistrophe ("turn/counter-turn") structure of the verses. The metrical form of the lyrics in each strophe (verse) was metrically and probably musically replicated by its corresponding antistrophe. The narrative content of the antistrophe also tended to reflect the strophe it mirrored. Additionally, tragic laments often alternated between two distinct voices responding to each other, repeating phrases and meters. These musical echoes would have acted as recurring emotional themes and may have also been reflective of the music of real funerary practice in Athenian culture.

One musical motif with strong associations with tragedy and lamentation is the call of the nightingale, which is found written as *"Itun, Itun"* – but probably sounded more like "itoo itoo" when sung. This repetitive sound is described in Sophocles' *Electra* as "the grief-bewildered bird" as Electra calls out the refrain (147–149). The chorus of Aeschylus's *Suppliant Women* illustrate musical evaluative conditioning when they sing that if anyone hears their lament he will think he hears the voice of the nightingale, who they describe as the "sad wife of Tereus" (57–72). This auditory link between the song of the nightingale and lamentation is very old: our earliest reference is a beautiful passage in *The Odyssey* (19, 518–523) where Penelope likens her grief over her husband's absence to the "many-toned sound" of the nightingale mourning for her dead son, Itys.

According to legend, the nightingale was once an Athenian princess named Procne who married Tereus, a foreign king. Tereus raped Procne's sister, Philomela, and tore out her tongue so she could not reveal the crime. But Procne found out and in a fit of revenge, killed their son, Itys. Procne and Philomela tried to escape, pursued by Tereus until the gods transformed them all to birds – the unintelligible Philomela into the swallow with its babbling birdsong; Tereus into the hoopoe with its long beak indicating the sword he tries to use to kill the sisters; and Procne to the nightingale always calling in deep regret for her dead son "Itys! Itys!" The real birdsong of the nightingale is intricate and trilling, and the epithet "many-toned" is quite apt. The bird's frenetic movements and trembling throat combined with its trilling call evoke the emotionality of lamentation.[33] In Greek culture, the bird was known as "the songstress" and was well known for her distinctive call. In tragedy, the song of the nightingale became associated with a measured sorrowful moan connected to female mourning as opposed to a sound that would provoke a sudden brainstem reflex, like a scream or piercing cry. For example, the male chorus in Sophocles' *Ajax* sings that the mother of Ajax, when learning of her son's death, will "howl with grief"; this will not be "a melancholic lament, not the Nightingale's soft song" (624–634).[34]

References to the nightingale evoke the mournful lament of memorialized pain, not the sudden and more aggressive outpouring of grief associated with new suffering. There is one such unsettling moment in the play, *Rhesus* (546–560) when the Trojan watch is roused by a sound and knows right away it is the sound of the nightingale "singing of her murderous marriage, with her *many-toned* voice."[35] The nightingale was a bird of the night, but here the song also creates a sense of foreboding of the grief that is to come in the Trojan War. In Euripides' *Helen*, when the chorus wants to sing a sad song for the suffering of Helen and the Trojans, they call on the "tearful nightingale" to "come, trilling with quivering beak, to help in my lament" (1107–1116), and one of the most famous examples is the elders of Argos in Aeschylus's *Agamemnon*, who compare the ill-fated Trojan princess, Cassandra, to the nightingale – the distinctive sound of its birdsong evoking the sorrow of a captive girl about to be killed:

> *You are mad! Possessed by some god,*
> *Singing this discordant song for yourself.*
> *You are like a nightingale, always calling,*
> *A mind full of misery, mourning, calling*
> *"Itys, Itys,"*
> *Surrounded by sorrow, a life steeped in grief.*[36]
> Aeschylus, *Agamemnon*, 1140–1145

Ancient drama is replete with these kinds of lyric motifs, and by considering evaluative conditioning we can understand them as powerfully communicative culture markers that inform emotional response, rather than only literary devices or sophisticated poetic allegories. They play a distinct and important role in communicating emotions and eliciting a shared response from the audience who knew them well. The bottom-up sensorial information generated by such evaluative conditioned verbal utterances are capable of producing powerful visual imagery and activating top-down stored memories as articulated by Sophocles' Electra:

> *Could anyone forget*
> *The horror of a parent's death?*
> *Or learn to be silent as a baby?*
> *My mind is stamped in the image of a crying bird,*
> *Who calls for her dead child forever, "Itys, Itys!"*
> *And mindless with grief, she is Zeus' messenger.*
> Sophocles' *Electra*, 145–149[37]

A somatosensory analysis of Aeschylus

Poetic language, especially when conceived for live performance, conveys a large amount of sensory information, what Starr has described as "the subjective experience of sensation without corresponding sensory input."[38] Such language that describes or sounds like a sensory experience activates the same areas of the brain associated with tangible sensory inputs and the processing of memory.[39] To briefly illustrate this point, I take two examples from Aeschylus's *Agamemnon*. Aeschylean poetic language is replete with words that communicate a multiplicity of meanings and metaphorical phrases that operate quite differently when uttered live than they do when read on the page. One notable example occurs at *Agamemnon* 612 when Clytemnestra is speaking to Agamemnon's Messenger. She sends him back to her husband with the message of her own regarding her fidelity, and says that she "knows as much about the pleasures of another man as she does of *steeping* metal."

In Greek the word for "steeping" is *baphas* – a term that would have evoked a specific set of sense memories with important ritualistic connotations for the audience. The word means to "dye" fabric or "temper" metal. When spoken it sounds like the very thing it describes – a plosive sound like the submerging of a swath of fabric into a vat of liquid dye, or the plunging of a hot piece of metal into cold water. The English word *baptize* is derived from it. But *baphas* can also be used as a noun and applied to a dyed garment, particularly one of ritual import. Noticeably, Aeschylus uses it to describe the crimson-dyed tapestries that Agamemnon walks on to enter his house (*Agamemnon* 960), the robe that enmeshed him in death (*Libation Bearers* 1013), and the saffron-dyed veil of Iphigenia that is imagined by the chorus as slipping from her head at the moment of her death (*Agamemnon* 238). This last instance connects Clytemnestra's use of *baphas* at 612 with the *baphas* of Iphigenia. At 238 the chorus describe the moment of Iphigenia's death thus:

κρόκου βαφὰς δ' ἐς πέδον χέουσα
The saffron veil poured onto the ground.

Compare this line to Clytemnestra's words to the messenger at 612:

ἄλλου πρὸς ἀνδρὸς μᾶλλον ἢ χαλκοῦ βαφάς.
Of another man, than of steeping metal (bronze).

The key terms here are *krokou baphas* ("saffron veil") and *chalkou baphas* ("steeping metal"). When Clytemnestra addresses the Herald, the audience has not heard the distinctive word *baphas* since the story of the sacrifice of Iphigenia told by the chorus during their entrance song. Yet, the phrase takes on another sensory dimension when we consider that the chorus sings how Iphigenia was killed at *Chalkis* (190) – the name meaning something like "metal" or "bronze town" – and that Clytemnestra uses the word *chalkou* for "metal." Thus, a sensory interpretation of the associations between the heard words signify a deeper and more highly charged connotation within Clytemnestra's claim of fidelity that means something like "I measure my fidelity based on the events at Chalkis." Yet Clytemnestra's phrase does not only look back to evoke the murder of her daughter, but it also anticipates that metal ax that she will plunge into Agamemnon – a *baphas* of blood.

Aeschylus was evidently famous for these kinds of linguistic sensorial flourishes, leading Aristophanes to describe his poetry as *kōdōnophalaropōlous*, a multisensory term that means "with clattering horse bronzes" (*Frogs* 963). In performance, these kinds of sensory associations in Aeschylean dramatic speech and song would also have been supported and enhanced by the rhythm, pitch, and tonal qualities of music that accompanied it. Let's dig a little deeper into Aeschylean somatosensory language and look at the opening scene of *Agamemnon*. The play begins with a solitary Watchman situated on the roof of the *skene*, which he describes as the House of Atreus. He tells us that he has been stationed there every night for an entire year to keep a look out for a signal beacon indicating that Troy has fallen to Agamemnon:

Gods! Free me from these labors!
I've spent a whole year up here, watching,
propped up on my elbows on the roof
of this house of Atreus, like some dog.
How well I've come to know night's congregation of stars,
the blazing monarchs of the sky, those that bring winter
and those that bring summer to us mortals.
I know just when they rise and when they set.
So I watch, watch for the signal pyre,
the burning flame that will tell us, Troy is taken!
Aeschylus, *Agamemnon* 1–10

These words were probably not set to music and were delivered in iambic trimeters, a heightened version of normal speech that is still song-like as Greek was a tonal language with the rise and fall of the pitch giving it a lilting "sing-song" quality. The translated phrase "My master's luck is mine" is an example of an iambic trimeter in English as it has three metric feet, each one made up of one short followed by one long syllable – de-dum/de-dum/de-dum. By way of contrast, when the chorus enter and first sing at line 40, we can see in the Greek that the meter has made a marked shift to a marching anapest (de-de-dum/de-de-dum), a driving beat that indicates that the chorus is moving into the orchestra. Other metrical shifts included moments where past events are recalled and sung of in a meter similar to Homeric epic verse (hexameter), and emotional moments, which were often set to lyrics in a trochaic tetrameter (dum-de-dum-dum).

The Watchman's very first word – *theos* – "Gods!" is a sudden and dramatic call to attention: in a theatrical environment without dimming houselights or a curtain to indicate the start of the performance, the Watchman's appeal to the heavens and the appearance of his masked face, presumably accompanied by a large gesture of appeal, would have focused the audience's attention directly on the *skene* roof, where both time (night at the end of the Trojan War) and place (The House of Atreus in Argos) are very quickly established. Now the audience knows they will be experiencing a *nostos* (homecoming) story of Agamemnon. With these important narrative elements established, Aeschylus then deploys a good deal of sensorimotor concepts to draw his audience into the experiential world of the Watchman. For example, at line 3 the Watchman tells us that he is "propped up on his elbows like some dog." This line, and the Watchman's posture, would have activated the sensorimotor regions of the brain in the spectator. In simplistic terms this causes the kind of "mirroring effect" described in Chapter Four whereby the sensorimotor properties of the words create a cognitive understanding of the discomfort of the Watchman. I am not suggesting that the audience members copied the Watchman's posture, but many of them would have felt a tinge of unease, shifted their position, and felt a degree of sensory empathy to his discomfort.

A 2013 fMRI study can help us understand how such speech descriptions affect us. In it blood oxygen levels were measured in the brain in certain key areas known for processing motor localizing tasks such as hand movements: the primary motor and sensory cortex and the lateral anterior temporal lobe, which are also associated with semantic processing.[40] The study used a series of phrases, which were classed as *Literal* ("The instructor is grasping the steering wheel very tightly"), *Metaphor* ("The congress is grasping the state of affairs"), *Idiom* ("The congress is grasping at straws in the crisis"), and *Abstract* ("The congress is causing a big trade deficit again"). The researchers explained their choices thus: "The literal sentences described physical actions. Metaphor sentences used the same action verbs metaphorically. The Idiom sentences used the action verbs in a highly conventionalized figurative manner. Finally, the Abstract sentences used verbs that had relatively low associations with actions."

The study concluded that the involvement of sensorimotor areas of the brain decreases as the level of abstraction increases. Therefore, the presence of an action verb alone is not enough to activate the brain's sensorimotor systems; instead, it is the use of that word in context as part of the semantic processing of "chunks" of language rather than the meaning of individual words. Another study used PET scans to show how people tend to process information about objects in a sensory manner and that the brain responds to objects such as tools in terms of how we manipulate them rather than a semantic notion of what the tool is used for.[41] For example, a traffic light was associated with its action as stopping movement rather than its function of moderating the flow of traffic, and a hammer with how it is held rather than the action of driving a nail. What these studies do for our purposes is establish a neural basis for a kind of linguistic kinesthetic empathy – rather than the action of watching another person engaged in movement, the same sensorimotor areas of the brain are activated by representations of movement implied by speech.

We can examine the Watchman's words by applying this neural model. "Gods! Free me from these labors!" is an abstract line, because we have no idea yet who this person is and what his labors are, and he appeals to heaven. "I've spent a whole year up here, watching" is very literal and provokes our sensorimotor systems instantly – we are not primarily concerned with the task or the time it has taken him but the expression of drudgery and weariness. In Greek the line reads as follows: *phouras eteias mêkos, en koimômenos* – the word order being: "Watching a whole year, laying." The last word on this line, *koimômenos*, means to guard at night but also implies to sleep or be lulled. The repeated sibilance of *phouras eteias mêkos, en koimômenos* and the wide-open vowels emphasize the sense of sleep and weariness. Then he says: "On this house of Atreus, propped up on my elbows like some dog." This is pure metaphor, but as the Rutvik study suggests, metaphoric language also registers strong activation in the brain's sensorimotor regions. We feel something of his discomfort. Now the language moves into the realm of the abstract as the Watchman fixates on the stars in the night sky and calls them the "congregation of stars" and the "blazing monarchs of the sky." For a moment the Watchman's physical weariness and discomfort are replaced by an abstract celestial image, abstract in language and meaning because the performance is taking place during the day and the Watchman is directing his spectators up into ambient extrapersonal space. The sensorimotor dynamism of the Watchman's first few lines offer us a plethora of information: time period – the end of the Trojan War; time of day – nighttime; place – the house of Atreus in Argos; state of mind – frustration; length of labor – one year; status – low; function – to watch; bodily state – uncomfortable; physical state – exhausted; solace – knowledge of the stars; and his first word acts to force attention and create an instant frame for performance.

Further intense somatosensory linguistic information is provided in lines 12–15 when the Watchman describes his "sopping bed" where he lies "tossing and turning" all night. Fear sits by his side, keeping him awake, and he desperately wishes he could just close his eyes and sleep. The alliteration of the beta sound surrounded by long vowels and the cadence of the meter create a sensory experience of the Watchman's exhaustion that goes beyond a narrative description of his situation:

To mê bebaiôos blephara sumbalein hupnô
I wish I could just close my eyes and sleep.

Just as the Watchman's speech is packed with emotional shifts, so the somatosensory information conveyed by the words and meter created an embodied experience for the audience. When at line 34 he implausibly imagines his king shaking his hand in welcome, the Watchman pulls his audience directly into his peripersonal space and activates the sensorimotor systems of their brains connected with the simple but intimate act of shaking a hand. Even things unspoken are rendered in somatosensory terms when the Watchman abruptly concludes his speech, presumably because either he sees the chorus approaching or Clytemnestra is emerging. Here he uses the vivid idiom "a great ox is standing on my tongue," and his words are suddenly halted. This is a superbly crafted and intimate portrayal of a man whose simple act of speech carried us to ancient Argos, the heavens, Troy, a sacred celebration, and a heartfelt act of welcome and made his audience feel for his frustration, weariness, wonder, fear, dejection, excitement, celebration, wish-fulfillment, and finally his silence.

Understanding the unintelligible

Oimoi! Iou! Papai! Pheu! Omoi! Attatai! Aiai! E-e! Oi! Ie! Io! Otototoi! Iouiou! Within the texts of Greek drama we can read these and other intense cries of pain, desperate appeals for help, exultant victory songs, delighted yelps, passionate calls to the gods, screams at the moment of death, and of course songs of lamentation and deep distress.[42] These "cries" are so difficult to translate into modern English and can seem so alien in performance, and yet their very sound acting upon the human auditory system and emotional processing centers was an integral element of the entire dramatic experience. Greek drama is replete with the vocalization of virtually unintelligible sounds that are uttered at the heights of emotion. In this final section I want to briefly examine the cognitive and affective power of this kind of unintelligible speech.

Andrew Ford has made a compelling argument that the various names and cult titles of Dionysos (*Bakchos, Iakchos, Lyaios, Euios, Bromios*, and many others) contain an "awareness that the roots of his name lay in meaningless vociferation."[43] *Bromios* means "lord of noise," and *Euios* and *Iakchos* are derived from the ritual inarticulate and emotionally expressive cries of "euoi!" and "iakche!" This is why the ancient Greek lyric poet Pindar named Dionysos "the ivy-knowing god, whom we mortals call Roarer and Shouter" (*Dithyramb for the Athenians*, Fr.75.9–10).[44] Ford points out that in Sophocles' *Antigone* the chorus describe Dionysos presiding over the "night calls" of the Bacchants who dance in frenzy in honor of "their lord Iakchos" (1146–1152). Ford adds, "this is speech that is not so much heard as witnessed . . . it seems more important that the god's *epikleseis* (invocations) be ritually performed than that they be understood."[45]

The idea of Dionysos presiding over emotionally intense unintelligible utterances and causing his followers to "forget their minds" encapsulates a good deal of what made up the theatrical experience for the ancient Athenians. The various cult practices of Dionysos exemplified the idea of engagement with the irrational side of human nature, and these feelings were provoked by the rhythmic entrainment of collective dance; the bodily dissonance that came from wearing and viewing masks; the witnessing of narratives that displayed excessive and outrageous (*hubristic*) behavior; the consumption of alcohol – itself a mind-altering substance – and, of course, the effect of otherworldly, distinct, and evocative music played on the *aulos*, accompanied by the beating of feet, the banging of timpani, and the often unintelligible but nevertheless emotionally distinct sounds produced by the human voice.

As students of ancient drama we are concerned with making the surviving texts of the plays understandable, but we can never come close to understanding the affective power of Greek drama without considering the emotional power of the unintelligible. As W. H. Auden commented about sung narrative in opera, an art form that initially sought to re-create the musical, visual, and narrative experience of Greek tragedy: "no good opera plot can be sensible, for people do not sing when they are feeling sensible."[46] At its most emotional moments, the lyrics of opera are often unintelligible. With this point in mind, Paul Robinson has argued that librettos of opera cannot be studied in isolation from the music they were conceived for and states that "opera singing . . . represents the most intractable enemy of intelligibility." This is because "composers write music for them (the singers) that can be produced only through certain specialized physical procedures, one of whose effects is that intelligibility must, in varying degrees be sacrificed."[47] Robinson sums up his point on the unintelligibility of operatic words by stating that an operatic text "really has no meaning worth talking about except as it is transformed by music."[48] Michel Poizat goes further and describes a sense of

> radical antagonism between letting yourself be swept away by the emotion (of the music) and applying yourself to the meaning of each word as it is sung . . . textual intelligibility and meaningless comprehension do not contribute to the production of emotion; in fact, they limit it or cancel it altogether.[49]

As an example, consider the last lines of Aeschylus's *Persians* (1066–1076): the chorus of Persian elders and the defeated king Xerxes jointly lament their loss at the hands of the Greeks. Even without knowing Greek, one can discern the emotional intensity of this song from the text. Here I print the Greek, a transliteration beneath, and then the translation by Janet Lembke and John Herington.[50] Read aloud the Greek or the transliteration:

Xerxes

βόα νυν ἀντίδουπά μοι.
(Boa nun antidoupa moi)
Din back my howling, my thumping

Chorus

οἰοῖ οἰοῖ.
(oioi oioi.)
Thousands and thousands

Xerxes

αἰακτὸς ἐς δόμους κίε.
(aiaktos es domous kie.)
Lament as you go to your houses

Chorus

ἰὼ ἰώ, Περσὶς αἶα δύσβατος.
(iō iō, Persis aia dusbatos.)
Sorry sorrow
Hard now to tread Persia's downtrodden Earth.

Xerxes

ἰωὰ δὴ κατ᾽ ἄστυ.
(iōa dē kat' astu.)
Wail as you step through the city

Chorus

ἰωὰ δῆτα, ναὶ ναί.
(iōa dēta, nai nai.)
Wailing wails, weeping

Xerxes

γοᾶσθ᾽ ἀβροβάται.
(goasth' habrobatai.)
Tread soft as you sob out your dirges

Chorus

ἰὼ ἰώ, Περσὶς αἶα δύσβατος.
(iō iō, Persis a-ia dusbatos.)
Sorrow our sorrow,
    Hard now to tread Persia's downtrodden Earth.

Xerxes

ἰὴ ἰὴ τρισκάλμοισιν,
ἰὴ ἰή, βάρισιν ὀλόμενοι.
(iē iē triskalmoisin,
iē iē, barsis olomenoi.)
Mourn mourn
    the men in the ships   three-tiered ships

Mourn mourn
>   your sons dead and gone   dead and gone

Chorus

πέμψω τοί σε δυσθρόοις γόοις.
(pempsoo toi se dusthroo-is goois.)
To slowdinning dirges we shall lead you home.

These utterances would have sounded quite alien and extreme to the original Greek audience in 472 BCE, but they represent a common trope in tragedy: the unintelligible cry.[51] It is striking just how many long vowel sounds appear in these 11 lines. The final word is the plural form of the Greek word *goos*, meaning to weep or wail. The word is predominantly all vowel sounds, with the double "o" sound dominating the soft "g" and diminutive "s". Vowels are significant communicators of human affective states and originate from the vibration of the vocal folds. The quality of the vowel sound (closed – "oo", open – "ah", or mid – "oh") is formed by the mouth. Alternatively, consonants are created by the exhalation of air and by the mouth. Singers must usually emphasize vowel sounds and diminish consonants or too much air will be expelled and the throat and mouth muscles will become tense. This can result in vocal strain, choked sounds, off-pitch notes, and vocal "cracking"; as the singer becomes higher and louder, these problems can get worse. The vocal technique known as *legato* was developed to deal with this. *Legato* elongates the vowels to carry the melody and shortens and diminishes the articulation of consonants, which can be another contributing factor in the unintelligibility of sung lyrics.[52] This technique relaxes the acoustic chambers of the throat and mouth and increases volume by resonance.

It sounds counter-intuitive that unintelligibility can actually increase emotional understanding and empathy, but physical vocal alterations, like the ones described above, which are used to communicate recognizable vocal patterns, can elicit sympathetic emotional responses in listeners. Human vocalization transmits affective information by tightening and loosening the vocal folds of the larynx to raise and lower pitch, by placing the larynx higher or lower in the throat to shorten or lengthen the resonating cavity to create distinguishable patterns, and by shaping the mouth to create recognizable vowel sounds and consonants. For example, a singer's lowered larynx can induce a sympathetic lowered larynx in listeners, who may feel like they "have a lump in their throat." This can be replicated in a simple experiment before any kind of audience: when speaking to them, lower your voice markedly and say something serious in a grave tone, then ask how many people in the audience swallowed or felt their vocal chords lower. Then start speaking in a high-pitched excited voice and note how many people feel tightness in their own larynx. The voice can provoke marked bodily reactions in others – another form of powerful kinesthetic empathy. Emotive vocalizations can also affect the listener's neurobiological responses. For example, the sound of somebody screaming is associated with the production of adrenaline, and this can induce an adrenaline-heightened state in the listener – just listen to the film music from *Jaws* or *Psycho* alone in a dark room.[53]

These basic forms of vocal emotional communication seem to be innate in the functional auditory channels of humans. Snowden and Teie have described how the human fetus can hear at 24 weeks and is therefore exposed to 4 months of constant sound *in utero*. The sound of the mother's heartbeat has been estimated at 25 decibels above the baseline ambient noise in the womb. This is by far the loudest noise the fetus can hear. Another prominent sound at this point of human development is the maternal voice, which has been calculated as being four times louder *in utero*. Higher frequencies are far more perceptible because of the tissue that surrounds the womb and absorbs the lower frequencies. This makes consonants very difficult, if not impossible, to discern.[54] What can be heard are the pitch variations of the mother's vowel sounds – the melody of her speech. These melodies are described as "harmonically consonant," which is another reason why distinct cultural groups all report a strong dislike of dissonant music.

Snowden and Teie point out that it is "the prosody of languages that forms the basis of music." Therefore, there are discernible variations in the traditional music of different cultures depending on the prosody of the native language. Newborns will mimic the timbre of their mother's voices in their early cries. For example, research has shown that French babies show a marked preference for spoken French even when a recording is played with all the discernible consonants and vowel sounds removed, leaving only the melody. Studies have also found marked differences between the rhythms of French and English music and that these mirror the respective rhythms of French and English spoken language.[55]

The effect of Aeschylus's Persian lament on his audience was profound. We know this because, despite only having one known performance at the City Dionysia in 472 BCE, the highly emotive vowel sounds were remembered and passed on by enough of the audience (or re-performed) that nearly 70 years later Aristophanes can make a comic comment on them in *Frogs*. During the competition between Euripides and Aeschylus in the underworld, Aeschylus mentions how proud he was of *The Persians* and Dionysos replies:

> I really loved it when we heard the dead Darius and right then
> The chorus struck their hands like this and went "iaow-oy!"
> (1028–1029)

The audience must have known the reference or the joke would have fallen flat. Even in comic form Dionysos affirms his propensity for the accentuation of heightened emotionality and the prevalence of sound over sense in the works performed in his honor.

This chapter has only scratched the surface of the somatosensory effects of the words of ancient drama, and yet even through this briefest of surveys we can discern the affective qualities of the text in its original incarnation as speech and song in performance. The sounds of the words, the rhythm of the meters, the use of the mask, and the visceral sounds of the unintelligible contributed greatly to overall emotionality of the work. In the final chapter we will pull all of the strands of this study together and see how the entire theatrical experience blended dissonance, dissociation, cognitive absorption, and empathy and show how the narratives of the plays that we can read in texts today were conveyed by a richly affective performative environment

Notes

1 Nietzsche 1993: 103. Walton has called the focus on just the words of ancient drama "absurd." He puts this down to a paucity of information on how the plays were staged and the analysis of Aristotle's *Poetics* as favoring plot construction over other important theatrical values. Walton 2015: 8.
2 Hall 2002: 3.
3 Brattico et al. 2010: 308.
4 McGurk and MacDonald 1976: 746–748.
5 For a digital video demonstration of the McGurk effect recorded by the BBC program *Horizon*, Try the McGurk Effect! Is Seeing Believing? BBC Two. "Is Seeing Believing" (2010–2011), see www.youtube.com/watch?v=G-lN8vWm3m0.
6 Besson et al. 2015:88.
7 Zatorre et al. 2007: 547–558.
8 Goldhill 2012: 90; Nooter 2012: 12.
9 Juslin and Västfjäll 2008.
10 Zatorre et al. 2007: 555.
11 Cross 2008: 147–167.
12 *Encomium of Helen*, 9.
13 Pinker 1997: 534.
14 Obermeier et al. 2012.
15 Salimpoor et al. 2011: 257–262.
16 See Kringelbach et al. 2008.
17 See Kringelbach et al. 2008.
18 Cherry 1953: 975–979.
19 Moray 1959: 56–60.
20 Driver 2001: 53–78.
21 Treisman 1969: 282–299.
22 Mesgarani and Chang 2012: 233–236.
23 Ross et al. 2007: 1147–1153.
24 Schall and von Kriegstein 2014.
25 Kim and Davis 2011.
26 Csapo and Slater 1995: 139.
27 Taplin 2001.
28 Hall 2000: 26.
29 Bentley 1952.
30 Wu et al. 2014: 1–7.
31 Starr 2010: 276–279.
32 For a description of the various modes see West 1992: 177–189.
33 Steiner 2013: 181 n. 27.
34 In Plato's *Phaedo* (58a), Socrates says that no birdsong should be associated with the lament, not even "the nightingale, the swallow or the hoopoe." See Chandler 1934: 78–84; Steiner 2013: 173–208.
35 The word is *Polychordotatia* – the same term used by Plato to describe the aulos (*Republic*, 399d).
36 For a comic evocation of the sound of the nightingale, see Aristophanes' *Birds* 209–222. Even here her birdsong is associated with plaintive lamentation.
37 Woodruff 2009.

38 Starr 2010: 276.
39 On neuroaesthetic approaches to poetry, see Burke 2014; Jacobs 2016 and 2015; Marin 2014; Conrad 2015: 40–42; Starr 2013.
40 Desai et al. 2013: 862–869.
41 Kellenbach et al. 2003: 30–46.
42 For a comprehensive list, see Nordgren 2015: 74–187.
43 Ford 2011: 345.
44 Ford 2011: 344.
45 Ford 2011: 347.
46 Auden 1961.
47 Robinson 2002:36.
48 Robinson 2002:46.
49 Poizat 1992: 36. I am grateful to an innovative doctoral dissertation by Carlo Zuccarini (2012) for leading me to scholarship on opera and unintelligibility. He adds, "some of the most powerful emotions that are experienced by listeners occur precisely when the words become unintelligible, namely when the voice becomes music and loses its connection with language, thereby destroying language."
50 Lembke and Herington 1981: 92–93.
51 Hall 1989: 83.
52 Zuccarini 2012: 23–24.
53 Snowdon and Teie 2013: 133.
54 Huotilainen 2013: 102–103.
55 Patel and Daniele 2003.

Bibliography

Auden, W.H. 1961. "American Christmas" in *Time* LXXVIII: 26.

Bentley, E. (Ed.). 1952. *Naked masks: Five plays*. New York: EP Dutton.

Besson, M., Chobert, J., François, C., Astésano, C. and Marie, C. 2015. "Influence of musical expertise on the perception of pitch, duration and intensity variations in speech and harmonic sounds" in Astesano, C. (Ed.), *Neuropsycholinguistic perspectives on language cognition: Essays in honour of Jean-Luc Nespoulous*. London/New York: Psychology Press: 88–102.

Brattico, E., Alluri, V., Bogert, B., Jacobsen, T., Vartiainen, N., Nieminen, S. and Tervaniemi, M. 2010. "A functional MRI study of happy and sad emotions in music with and without lyrics" in *Frontiers in Psychology* 2.308: 1–16.

Burke, M. 2014. "The neuroaesthetics of prose fiction: Pitfalls, parameters and prospects" in *Frontiers in Human Neuroscience* 9.442: 1–6.

Chandler, A.R. 1934. "The nightingale in Greek and Latin poetry" in *The Classical Journal* 30.2: 78–84.

Cherry, C. 1953. "Cocktail party problem" in *Journal of the Acoustical Society of America* 25: 975–979.

Conrad, M. 2015. "On the role of language from basic to cultural modulation of affect: Comment on 'The quartet theory of human emotions: An integrative and neurofunctional model' by S. Koelsch et al." in *Physics of Life Reviews* 13: 40–42.

Cross, I. 2008. "Musicality and the human capacity for culture" in *Musicae Scientiae* 12.1 (suppl): 147–167.

Csapo, E. and Slater, W.J. 1995. *The context of ancient drama*. Ann Arbor, MI: University of Michigan Press.

Desai, R.H., Conant, L.L., Binder, J.R., Park, H. and Seidenberg, M.S. 2013. "A piece of the action: Modulation of sensory-motor regions by action idioms and metaphors" in *NeuroImage* 83: 862–869.

Driver, J. 2001. "A selective review of selective attention research from the past century" in *British Journal of Psychology* 92.1: 53–78.

Ford, A. 2011. "Dionysus' many names in Aristophanes' frogs" in Schlesier, R. (Ed.), *A different god? Dionysos and ancient polytheism.* Berlin: De Gruyter: 343–356.

Goldhill, S. 2012. *Sophocles and the language of tragedy.* Oxford/New York: Oxford University Press.

Hall, E. 1989. *Inventing the Barbarian.* Oxford: Oxford University Press.

Hall, E. 2002. "The singing actors of antiquity" in Easterling, P. and Hall, E. (Eds.), *Greek and Roman actors: Aspects of an ancient profession.* Cambridge: Cambridge University Press: 3–38.

Hall, P. 2000. *Exposed by the mask.* New York: Theater Communications Group.

Huotilainen, M. 2013. "A new dimension on foetal language learning" in *Acta Paediatrica* 102.2: 102–103.

Jacobs, A.M. 2015. "Neurocognitive poetics: Methods and models for investigating the neuronal and cognitive-affective bases of literature reception" in *Frontiers in Human Neuroscience* 9: 186.

Jacobs, A.M. 2016. "The scientific study of literary experience" in *Scientific Study of Literature* 5.2:139–170.

Juslin, P.N. and Västfjäll, D. 2008. "Emotional responses to music: The need to consider underlying mechanisms" in *Behavioral and Brain Sciences* 31.5: 559–575.

Kellenbach, M.L., Brett, M. and Patterson, K. 2003. "Actions speak louder than functions: The importance of manipulability and action in tool representation" in *Journal of Cognitive Neuroscience* 15.1: 30–46.

Kim, J. and Davis, C. 2011. "Testing audio-visual familiarity effects on speech perception in noise" in *International Congress of Phonetic Sciences. Hong Kong* 4: 1062–1065.

Kringelbach, M.L., Vuust, P. and Geake, J. 2008. "The pleasure of reading" in *Interdisciplinary Science Reviews* 33.4: 321–335.

Lembke, J. and Herington, C.J. (Trs.). 1981. *Aeschylus: Persians.* Oxford: Oxford University Press.

McGurk, H. and MacDonald, J. 1976. "Hearing lips and seeing voices" in *Nature* 264: 746–748.

Marin, M.M. 2014. "Crossing boundaries: Toward a general model of neuroaesthetics" in *Frontiers in Human Neuroscience* 9.443: 1–5.

Mesgarani, N. and Chang, E.F. 2012. "Selective cortical representation of attended speaker in multi-talker speech perception" in *Nature* 485.7397: 233–236.

Moray, N. 1959. "Attention in dichotic listening: Affective cues and the influence of instructions" in *Quarterly Journal of Experimental Psychology* 11.1: 56–60.

Nietzsche, F. 1993. *The birth of tragedy: Out of the spirit of music.* London: Penguin.

Nooter, S. 2012. *When heroes sing: Sophocles and the shifting soundscape of tragedy.* Cambridge: Cambridge University Press.

Nordgren, L. 2015. *Greek interjections: Syntax, semantics and pragmatics* (Vol. 273). Berlin: Walter de Gruyter GmbH & Co KG: 74–187.

Obermeier, C., Menninghaus, W., von Koppenfels, M., Raettig, T., Schmidt-Kassow, M., Otterbein, S. and Kotz, S.A. 2012. "Aesthetic and emotional effects of meter and rhyme in poetry" in *Frontiers in Psychology* 4.10: 1–10.

Patel, A.D. and Daniele, J.R. 2003. "An empirical comparison of rhythm in language and music" in *Cognition* 87.1: B35–B45.

Pinker, S. 1997. *How the mind works.* New York: Norton.

Poizat, M. 1992. *The angel's cry: Beyond the pleasure principle in opera.* Ithaca, NY: Cornell University Press.

Robinson, P. 2002. *Opera, sex and other vital matters*. Chicago: University of Chicago Press.

Ross, L.A., Saint-Amour, D., Leavitt, V.M., Javitt, D.C. and Foxe, J.J. 2007. "Do you see what I am saying? Exploring visual enhancement of speech comprehension in noisy environments" in *Cerebral Cortex* 17.5: 1147–1153.

Salimpoor, V.N., Benovoy, M., Larcher, K., Dagher, A. and Zatorre, R.J. 2011. "Anatomically distinct dopamine release during anticipation and experience of peak emotion to music" in *Nature Neuroscience* 14.2: 257–262.

Schall, S. and von Kriegstein, K. 2014. "Functional connectivity between face-movement and speech-intelligibility areas during auditory-only speech perception" in *PloS One* 9.1: e86325.

Snowdon, C.T. and Teie, D. 2013. "Emotional communication in monkeys: Music to their ears?" in Altenmüller, E., Schmidt, S. and Zimmermann, E. (Eds.), *The evolution of emotional communication: From sounds in nonhuman mammals to speech and music in man*. Oxford: Oxford University Press: 133–151.

Starr, G.G. 2010. "Multisensory imagery" in Zunshine, L. (Ed.), *Introduction to cognitive cultural studies*. Baltimore, MD: Johns Hopkins University Press: 276–279.

Steiner, D. 2013. "The Gorgons' lament: Auletics, poetics, and chorality in Pindar's Pythian 12" in *American Journal of Philology* 134.2: 173–208.

Taplin, O.P. 2001. "Masks in Greek tragedy and in Tantalus" in *Didaskalia* 5.2.

Treisman, A.M. 1969. "Strategies and models of selective attention" in *Psychological Review* 76.3: 282–299.

Walton, J.M. 2015. *The Greek sense of theatre: Tragedy and comedy*. London: Routledge.

West, M.L. 1992. *Ancient Greek music*. Oxford: Oxford University Press.

Woodruff, P. 2009. "Electra" in Meineck, P. and Woodruff, P. (Trs.), *Sophocles: Four plays*. Indianapolis/Cambridge: Hackett Publishing.

Wu, Z.M., Chen, M.L., Wu, X.H. and Li, L. 2014. "Interaction between auditory and motor systems in speech perception" in *Neuroscience Bulletin* 30.3: 490–496.

Zatorre, R.J., Chen, J.L. and Penhune, V.B. 2007. "When the brain plays music: Auditory – motor interactions in music perception and production" in *Nature Reviews Neuroscience* 8.7: 547–558.

Zuccarini, C. 2012. *Enjoying the operatic voice: A neuropsychoanalytic exploration of the operatic reception experience*, Doctoral dissertation, School of Social Sciences Theses.

7   *Metabasis*
Dissociation and democracy

Aristotle describes *metabasis* as a movement of fortune, such as a change in the plot or a moment of transformation, and this concept is central to his understanding of how ancient theatre operates. In this last chapter, I relate Aristotle's concept of *metabasis* or "transformation" to the generative predictive model of cognition to suggest that transformation was fundamental to what Greek drama was striving to achieve in performance. It was in fact through the experiential elements of drama I have described that the emotional and empathetic responses of the audience were enabled to transform.

In this final chapter I examine how the mental states of dissociation and cognitive absorption promoted by the performance of ancient drama enhanced decision making and empathy. I describe very recent advances in the field of neuroaesthetics on what has been called the brain's Default Mode Network (DMN), which has started to establish neural evidence for the cognitive advantages of absorption. I then explore the cognitive effects of mind-body dissonance and propose that Greek drama provoked this kind of cognitive state and demonstrate how empathy can be enhanced by dissociation. This advances my central argument, which is that the promotion of empathy was an imperative for theatre in ancient Athenian society, and I offer some examples of empathy in political action in Athenian society, suggesting a "spill-over effect" from the theatre into public political life. The book ends with a personal appeal regarding the ongoing political and cultural necessity of theatre in free societies, including the need for collective physical embodiment over individual virtual disconnection, and the continuing power of Greek theatre to offer people the ability to experience empathy.

The Default Mode Network

The process of cognitive absorption has been linked to the workings of the brain's Default Mode Network (DMN), a relatively new discovery in neuroscience and thus its function and physiology are not yet completely understood. Electroencephalogram (EEG) and fMRI research suggests that the network is made up of at least the brain's posterior cingulated cortex, which has been associated with introspection and self-monitoring; the ventromedial prefrontal cortex (consciousness and executive function); the inferior parietal lobe (emotional processing and sensory information); and the medial temporal lobe (processing of memories). The substantia nigra may also be connected to the working of the DMN. One part of the substantia nigra regulates the production of dopamine, while another transmits signals from the basal ganglia to other regions of the brain. This area is key for the processing of movement and planning and has a connection to the brain's reward system.

In a groundbreaking study, Starr, Rubin, and Vessel have shown that the DMN is activated when people have a strong response to an artwork, in this case paintings.[1] This was a surprising finding: the DMN is one of the brain's basic operating systems and is active when the mind is at wakeful rest. It produces less energy when the mind is distracted by an external stimulation and returns to baseline when a person is not focused on a specific task. The researchers fully expected that when the participants responded most strongly to an artwork, either positively or negatively, their DMN area would register the lowest activation as energy was diverted to the areas of the brain dealing with external stimuli. But this was not the case; in actuality the DMN was most active at the moments when the participants reported being most moved by an aesthetic experience.

The neural activity of the DMN may explain the type of cognitive absorption created by mimetic aesthetic events, such as music, poetry, innovative narrative action, and visual art, all essential elements of Greek drama. The DMN internalizes these stimuli, creating a profound and deeply personal emotional experience. It is as if we are jolted from our normal associative cognitive functions by the aesthetic work and its novelty, strangeness, dissonance, odd juxtapositions, different rhythms, engaging movements, and so on. Once the work has gotten our attention, the DMN kicks in to compare what it knows with what we are experiencing. Starr describes how these processes "happen in the space between what is relatively predictable . . . and what is perhaps, strangely, not."[2] This new work on the DMN provides fMRI evidence for the theory of predictive processing in that it shows that even when "at rest" our minds are wandering freely or daydreaming and at their most introspective and not consciously being provoked by external perceptual stimuli. Clark has surmised that this research "reflects the ceaseless activity of the neural machinery whose compulsive prediction-mongering maintains us in constant yet ever-changing states of expectation."[3] The DMN has also been linked to Theory of Mind and its role in empathy and socially induced cognitive emotion regulation.[4] This is when a person or persons provide a target person with alternative interpretations for emotion-triggering stimuli to change his or her emotional state.[5] In a nutshell, this is what the theatre does – it engages the audience members' empathetic pathways to care about the predicaments and situations of the actors they are watching, even though their performances are mimetic in nature. This new research on the DMN indicates that this kind of empathetic engagement is best provoked by surprisal and supported by cognitive absorption. A practical application of this research has shown up in recent studies, which have detected a decline in DMN activity in veterans diagnosed with posttraumatic stress disorder. It has therefore been proposed that cognitive absorption leading to activation of the DMN has tangible therapeutic benefits.[6]

We might momentarily relate this new research on modern combat veterans to what we know of life in fifth-century Athens, where the population was subjected to a time of almost constant war. I have argued elsewhere that one of the social functions of drama was to provide a place of communal therapy for a society traumatized by war. This notion of drama as therapy is also borne out by its relationship to the cult of the healing god Aesclepius, when Sophocles brought the cult to Athens, where it was established next to the Theatre of Dionysos.[7] This interpretation of one of the aspects of classical drama also makes sense of Aristotle's use of the term *catharsis* as a healing or purgation being induced by mimesis and producing emotional and empathetic responses in the audience. Cognitive absorption and the DMN may have been key neural elements in the efficacy of *catharsis*.[8]

En xena xenon – "a stranger in a strange land"[9]

One of the common factors in the absorbed state and central to the predictive error correction cascade is the cognitive processing of surprisal, a common cognitive factor in all of the experiential elements of Greek drama we have explored here. When we process an atypical situation, our emotional states are heightened and help motivate us to think and behave in ways that promote adaptation and enhance our potential for survival and success. The studies of Huang and Galinsky have demonstrated that as humans move beyond infancy we learn to display bodily expressions that contradict our feelings. They have described this as 'mind-body dissonance" and explored how it allows people to be less constrained by their normal mental framework, which can lead to cognitive expansion.[10] This is similar to Starr's DMN theory of aesthetics. Mind-body dissonance is itself an atypical state that causes a sense of temporary incoherence, which motivates us to "embrace atypical exemplars." This, in turn, helps evoke empathy as we come to accept novel ideas. Greek drama was just such a dissociative, dissonant experience that promoted cognitive absorption. Hence, in antiquity we find many references to "soul-moving" or "spell-binding" qualities of performance and its emotional effects, including negative attitudes from both Plato and Aristotle that it can sometimes be culturally disruptive and politically dangerous.[11] Another comes from Plutarch, who, in his rhetorical treatise *On the Fame of the Athenians*, posed the question "were the Athenians more famous in war than in wisdom?" Within the subsequent discussion he remarks on the incredible popularity and impact of Athenian tragedy in particular, which shows that it was "amazing" (*thaumatos*) and that it beguiled because the "mind is enthralled by the pleasures of language and becomes *anaisthetos*." The literal translation is "insensible," but perhaps what Plutarch is really trying to explain is tragedy's power of absorption (348c).

Plutarch goes on to ask what benefits drama brought Athens and famously remarks that the city spent more on theatre than it did on defense (348f–349a). He rather factiously concludes that the tripods awarded to the victorious Choregoi were nothing more than "an empty memorial of their vanished estates." But even this dismissive remark contains another important socio-political facet of drama – the glorification of aristocratic patrons within a democratic system, a means by which class tensions in the city were alleviated. What may have seemed senseless to Plutarch writing in the Roman world in the late first and early second centuries CE may well have been the very thing that made Athenian drama such a cultural and political force worthy of the kind of expenditure and state commitment he describes. That is its power to dissociate and absorb, and in turn, to expand the mindsets of those who experienced it, and thereby influence the decision-making capabilities of that audience. In this way dissociation and absorption (*anaisthētos* and *thaumatos*) can become political agents in that they have a marked cognitive impact on the way in which people make decisions, understand unfamiliar situations, and empathize with outsiders. According to a good deal of research on dissociation, the kind of decisions we make when we are most disorientated are, in many ways, among the very best.

Let me demonstrate this with a practical example: imagine waking up with a start in a strange, totally unfamiliar place, after hearing am alarming, loud, high-pitched noise. Immediately your survival mechanisms automatically prepare you for fight or flight – your heart rate races, your respirations elevate as you gulp in more oxygen to fuel your brain, and you prepare your muscles for action. Your pupils constrict because of the bright overhead light and your skin temperature increases as blood is rushed by the pumping of your heart to your extremities. Your endocrine system releases a flood of adrenaline to give instant additional energy to your cardiovascular system. Your senses quickly alert you to the fact that you are restrained by three straps across your body and something is gripping your upper right arm. As your eyes adjust to the bright light, you perceive and then process the features of a face staring at you. In this alarming situation your predictive cascade is forced to make incredibly quick decisions about any potential threat. Your first autonomic action might be to flinch and to pull your own head and body away to try to place distance between you and the potential threat.

Now your perceptive systems quickly gather sensory information that is compared to cognitive models stored in your brain's working memory. Noise – a repetitive beeping and the low sound of an engine; vision – lights, a bench, a man in uniform, a large orange bag; touch – a tightening on your left arm, straps loose around your body; smell – plastic, canned air and now speech – "Okay, okay, I just need to take your blood pressure." Now you predict that you are in an ambulance and you are a patient. Now there is an immediate prediction error correction from the instinct to take flight or to fight – what has happened? Are you injured? Where? How badly? Based on this error correction your top-down executive functions seek to update the sensory information. You look again at the person's face and see that it has a calm but focused expression. This person's touch is firm but measured and does not hurt, and a male voice (it could just as easily have been a female voice) says, "Hi, I'm John, an EMT. You are in an ambulance. You had a fall and were out for a while, welcome back. It's okay, we have you. We are taking you to the hospital."

In this moment your cognitive systems are not considering external events or what is outside of the ambulance; you are completely focused on your immediate environment and the information you need to survive. The decisions you make in these kinds of extremely stressful situations are devoid of any second-guessing; they are initially guided by pure survival emanating from an initial sense of complete dissociation, which forces us into a state of heightened cognitive absorption to collect as much sensory information in as short a time as possible. Now as we scan the face and body of the EMT our empathetic systems come into play as we attempt to predict the intentions of another person for good or for harm by the expression on his or her face and by body posture and movement.[12] These kinds of emergency scenarios quickly deploy every cognitive system we possess to evaluate the environmental situation and how best to avoid entropy in the guise of death. We are, in effect, Friston's fish – in that we rely on autonomic systems that maximize our sensory-gathering capabilities and our bodily responses to prolong our survival. During these times of dissociation, we become optimally absorbed, and our abilities to judge the intent of actions are at their very best. This is the deployment of empathy for our own survival.

You made the right decision in the back of that ambulance. You did not push the EMT away, rip off the straps, and jump out of the back doors in terror; you correctly predicted that this person was there to help you. But for all the mental benefits provided by cognitive absorption and kinesthetic empathy, you also made an error – you predicted that the EMT was smiling and relaxed and therefore your injuries were not that bad and this calmed you and relaxed your heart rate, slowed your breathing, and lowered your blood pressure. But you were wrong – your predictive system accepted a lie.

It's a fact of life that in order to function in society we need to constantly project tiny lies about the way we are really feeling. We do it all the time – smiling and feigning interest in another person's long-winded story, pretending to be truly amazed by a friend's holiday photos. Acting stern and concerned for the benefit of a child who needs to be rebuked even though the offense is really very funny. The truth is that if we actually displayed our true emotional states as we experienced them, we would quickly alienate most of our friends, find our working environments very challenging, and have a very hard time motivating others or getting them to appreciate our concerns. The failure to recognize the affective states of others and moderate one's behavior as a result can be debilitating to some such as many people diagnosed with autism.[13]

In his comments on Athenian tragedy, Plutarch cites Gorgias: "he who deceives is more honest that he who does not deceive, and he who is deceived is wiser than he who is not deceived . . . for he has done what he promised to do" (348c). Huang and Galinsky's work on mind-body dissonance has shown that the time when a child learns to regulate his or her bodily expressions to aid the furtherance of interpersonal relations is widely regarded as a developmental marker of a level of social maturity. It is the development of empathy – the ability to adjust the outward markers of one's own feelings to accommodate the emotional state of another. When the EMT smiled, he was not expressing his own happiness, or laughing at something; he chose to wear that particular expression because he knew it would reassure his patient. The EMT may have actually been quite stressed at the situation, or agitated that a call came in right at the end of his shift, or perhaps the injury you had sustained was actually life threatening. Had you discerned any of those expressions you may have empathetically mirrored the EMT's expression, and this could have resulted in a dangerous spike in blood pressure, an anxiety attack resulting in shortness of breath and hyperventilation, or a sense of panic that might have hindered proper emergency care. Mind-body dissonance is vital to situational empathetic communication.[14]

This has ramifications for how we accept or reject atypical situations, such as those found in the narratives of all Greek dramas, which are nothing if not atypical. Topolinski and Strack have found that when a person's bodily expressions contradict their mental states, they will be more likely to embrace atypicality.[15] The EMT's smile is essentially a visual empathy signal: he has already decided not to display his true emotional state but to project a different expression all together for the benefit of somebody else. In doing this, our cognitive processes become more susceptible to different perspectives. For the EMT, this might mean that he does not rely on basic assumptions during his patient assessment, but looks beyond what he might normally expect to see and is more open to different explanations for what might have caused your fall and where else you might be injured. His own mind-body dissonant behavior – smiling when he is actually feeling quite concerned and tense – actually decreases his predictive processing fluency and promotes absorption, but also a search for predictions that are neither normal nor usual.[16] Processing fluency is our preferred cognitive state; most of us like surprises, but we prefer them infrequently. This is why many people will pay more for a brand-name product over the same product in a generic form, because the brand is familiar and predictable.[17] Disruptions to processing fluency can be disconcerting; for example, one study found that stocks did much better on the exchange after their initial public offering if they had a name that was easier to pronounce or remember when reduced to a three-letter ticker symbol. Thus, KAR was preferable to RDO because of ease of processing fluency.[18] Most of us prefer consistency in our perceptions, as the cognitive load required by generative error-correcting can also be debilitating and lead us away from acceptance of atypicality and toward its rejection. In Greek drama those moments of surprisal need to ultimately resolve into some sort of sense. Hence, Aristotle's words of wisdom in *Poetics*: "since the most amazing (*thaumatos*) incidents even among random events are those which we perceive to have happened as if it were on purpose" (1452a6–8).

Bodily expressions consistent with mental experience enhance emotional recognition and generate a sense of coherence, which increases processing frequency and accelerates decision making. This is a useful mechanism for most day-to-day cognitive decisions. If I want to turn left in my car, I know that if I turn the steering wheel to the left, the car will go in that direction. After I learn to do this and perform the operation many times, I barely need to think about it. However, if I now attach a trailer to my car and need to reverse using only my mirrors, it takes a great deal of practice to learn that I need to turn the wheel to the right to make the trailer reverse left. The first few times I try this maneuver I will take more time, be more cautious, and think more deeply. It is also notable that when we initially process these kinds of movements our faces tend to make strange smiling or grimacing expressions, or we bite our lips or tongues, attempting to suppress the fear we might really be feeling and instead "focus" on the difficult task in hand. This is self-produced mind-body dissonance, and whether performing it yourself or watching it in others, it has actually been shown to improve judgment.[19]

This is supported by a study by Rees et al. in which 76 participants were assigned three different emotional states: happy, sad, and ambivalent (both happy and sad). The researchers were looking to see if there was a correlation between mind-body dissonance as expressed by emotional ambivalence and the judgment of the participants.[20] They defined emotional ambivalence as the experience of strong emotions that pull people in different directions, for example taking pleasure in a touching human moment within a film that takes place in a wider tragic context, such as a war zone or concentration camp (their example was the 1997 film *Life is Beautiful*). This is akin to an audience taking pleasure from watching tragedy or experiencing sad music. They asked each group to write down a time when they had felt the emotion they had been assigned and then give their best estimate for the average temperature of eight geographically diverse American cities. What the study showed is that the group who had been asked to think about emotional ambivalence had a marked decrease in error from the happy and sad groups and an increase in receptiveness to alternative perspectives.

Maintaining two contradictory emotional states enhances an individual's perspective-taking tendency, in that he or she may be able to look at the world from the perspectives of others, interpret emotional states more accurately, and empathize with a greater variety of emotional states. One can derive pleasure from experiencing the performance of a tragedy that is expressing negative emotions and yet be moved by that very performance. This is a theatrical expression of mind-body dissonance, and it has been shown that these absorbed states can lead to greater feelings of empathy.[21]

I think these concepts are essential to understanding what Greek drama was trying to achieve: that is, to provide its audience with a dissociative and dissonant experience, which led to audience absorption, emotional contagion, and the acceptance of another perspective quite different from one's own. This premise makes sense when we place Greek drama within its original context as a state-run cult festival in honor of a god, who was known for his relationship to an altered state, liminality, and the merging of social and cultural boundaries. Other important Athenian cult practices also utilized dissociative cognitive practices, such as the Eleusian Mysteries, which disorientated its initiates with atypical somatosensory stimuli and disconcerting performances that included masked figures, invective poetry, and overwhelming visual effects. The Corybantic rites involved music and dance disorientating its devotees in a bid to affect a change in their mental states,[22] and Aristotle's concept of mimetic *catharsis* entails that one is transformed by the emotionality of the performance.[23]

Through dissonance and dissociation, provoked by the expert fusion of surprisal, empathetic projection, musical entrainment, expectancy and emotional affect, kinesthesia, and cultural memory, the theatre of Dionysos offered a gradual kind of *catharsis* in the form of empathetic understanding. This occurred under a dopa-mine-inducing sky that promoted receptivity toward alternative perspectives. In describing the cognitive and cultural benefits of the workings of the Default Mode Network – the neural architecture that supports the human predictive imagination – Starr has provided an apt description of what the Athenian theatre was trying to elicit in its audience, "with its continual reconfiguration of what we find valuable, [which] spurs us to seek new comparisons, new interpretations, new metaphors, new modes of understanding and perceiving."[24]

Dissociation – absorption – empathy

We can turn to recent neurological work on prediction and empathy to exam-ine how dissociation and surprisal, itself a form of momentary dissociation and absorption, operates to produce empathetic reactions. A 2015 study by Hein et al. examined how neural prediction error signals impacted learning between out-groups and, if mediated by positive emotions, enhanced empathy-related responses in the anterior insular cortex, as shown by fMRI imaging.[25] This brain region has been associated with decreased neural activity related to empathy def-icits concerning attitudes toward out-groups. In this study, participants received relief from pain inflicted on the hand by members of their own in-group and then again by members of a stated and identifiable out-group. The study assumed that expectations concerning out-groups are generally more negative than those of in-groups and that receiving help from an out-group member "is an unex-pected positive outcome that should elicit a strong positive prediction error." The results showed that the prediction error learning coupled with a positive affec-tive response did not only lead to more indications of empathetic response in the anterior insular cortex in response to out-group members stopping the partici-pant's pain, but also to other members of that same out-group observed having the same pain inflicted on them. This study showed that empathy can be quickly learned through prediction error correction and then transferred to others. This is one of the first mechanistic accounts of empathy in action, and it went on to predict reductions in empathy deficits with as few as two positive experiences with an out-group individual. The aim of the researchers was to provide evidence for the mechanisms that foster out-group empathy, so as to enhance intergroup contact for, as they correctly state, "deficits in empathy enhance conflict and human suffering."

This study used the sight of pain inflicted on another to measure the efficacy of predictive empathetic learning and seems to have detected some neural mecha-nisms that underpinned what Greek drama effectively provoked. This concept of learning empathy is ingrained in the idea of the function of the theatre in antiquity as a mimetic mechanism for education. The playwrights were called *didaskaloi* (teachers), and both Plato's and Aristotle's fascination with the role of perfor-mance to inspire good or bad behavior is a reflection of this this. Paul Cartledge has gone so far as to assert that the theatre was an essential part of the learning process of the Athenian citizen "to be an active participant in self-government by mass meeting and open debate between peers."[26] This echoes of Euripides in Aristophanes' *Frogs* (1009–1010), who exclaims that playwrights should be admired for "cleverness, and giving good advice since we improve the people in the cities." At the end of the play, Aeschylus is chosen to arise from the dead and save the city and to "educate the thoughtless" (1502–1503), for his "good ideas will bring great gain and spare us from the pain of war" (1529–1530).

There is contemporary evidence that empathy generated by mimetic means can have the kind of effect Aristophanes' chorus is wishing for. Malhotra and Liyanage showed how a four-day "peace workshop" in Sri Lanka fostered longer-lasting empathetic responses in out-group members than in a control group who were not exposed to the workshop.[27] This peace workshop involved participants from different groups who had been or were actively involved in conflict coming together for multiple days to explore and understand each other's perspectives. This included, shared tasks, story-telling, performance, dance, cooking and eating together, and team-building exercises based on building empathy. The participants were drawn from Sinhalese (mainly Buddhist), Tamil (mainly Hindu), and Muslim populations who had a long history of out-group antagonism and violence toward each other. The long-term results of the study of these peace camps showed a marked increase in reported empathy toward out-groups and an increase in donations to children's charities of out-groups as opposed to the control groups that did not attend the camps.

Closer to the mimetic impact of drama were the findings of a 2009 study in Rwanda, where xenophobic and bigoted radio broadcasts played a prominent role in inciting the violence between ethnic groups that resulted in the brutal genocide of more than 75% of the minority Tutsi population in three months in 1994, some 800,000 to 1 million people.[28] Radio is the most prominent form of mass communication in Rwanda, and this study by Elizabeth Paluck examined the impact on listeners of a new radio soap opera called *Musekeweya* ("New Dawn"), which tells the story of two unidentified ethnic groups, one of which is favored by the government and had grown more prosperous prior to being attacked by the other group. The soap opera deals with the aftermath of the violence and its effects on ordinary people, and it also dramatizes intergroup cooperation. Paluck established a control group who did not listen to the radio show and monitored eight groups representing a range of ethnic groups and geographic regions. These groups were monitored while listening to the radio broadcasts for one year; at the end of that period, within all the groups who had listened to the broadcasts there was a marked increase in empathy toward the out-group. Paluk noted that the radio program "did not change listeners' personal beliefs but did substantially influence listeners' perceptions of social norms. Normative perceptions were not empty abstractions but were realized by actual measured behavior, such as active negotiation, open expression about sensitive topics, and cooperation."[29] Mimesis can dramatically change minds, and empathy for out-groups can be learned, especially when learned via the predictive error-correcting of being absorbed by the twists and turns of a good dramatic plot.

Ancient empathy in action

This study could perhaps be accused of overemphasizing the empathy-inducing effects of ancient drama on the Athenians. After all, fifth-century Athens was a slave society, where women had no political rights and there was a state of almost-constant warfare. And yet, I see in these plays an attempt to provide the Athenians with an alternative view of the world around them and to view the festivals where they were produced as offering novelty, experimentation, surprise, and alterity. Plato's disdain for the fourth-century Athenian populace as a *theatrocracy* contains an element of truth about the social impact of drama: it can make the audience members feel that they have a right to freedom of expression and that their opinions and decisions have an impact on the direction of the state. Did this mean that drama was effective in helping to create a more empathetic and deliberative citizen body? I think so. But the kind of emotional impact generated by the theatre could work both ways: according to Plato, Socrates blamed Aristophanes' comedy *Clouds* for how he was perceived by the majority of the citizens who were now sitting in judgment on whether he should live or die.[30] But if the mythological Athenian citizen jury under guidance of Athena in Aeschylus's *Oresteia* could resolve the inter-generational bloodshed of the house of Atreus and even mollify the terrifying Furies, could the real jurors sitting in the audience not then be inspired to resolve the political unrest between the aristocrats and democrats in Athens in 458 BCE? Yet these plays can also seem overtly political and perhaps even one sided, such as Aristophanes' attacks of the populist demagogue Cleon, lampooning him directly in *Babylonians*, and when faced with a possible defamation suit presenting him as a slave in *Knights* and as a dog in *Wasps*. Yet even the comic plays of Aristophanes seem concerned with offering his audiences an alternative view of their current lives, often in absurdist and fantastical ways.

Aristophanes produced an overt appeal to the power of dramatic alterity to affect society at a time of great peril for Athens, when the state was facing total capitulation to the Spartans after years of debilitating war. In *Frogs*, staged in 405, he presented the god Dionysos heading down to Hades to bring back the recently deceased Euripides to "save the city." Though this is a comic premise to be sure, there must have been some sense that the dramas Athens produced were essential to the health of the Athenian democracy, which was now under existential threat. Aristophanes did not bring back Euripides, but rather Aeschylus, the grand old man of Athenian theatre and the famous Marathon fighter. His prediction seems to have been right, as Xenophon tells us that the Spartans spared the city in memory of its past victories over the Persians, though more contemporary geopolitical reasons may have been the real reason.[31]

In the fourth century, Athens, which was no longer a military power, gained new prominence as a center of Greek culture. The Theatre of Dionysos was enlarged and rebuilt in stone, and the plays of the three great fifth-century playwrights were written down and preserved in a state library as permanent markers of Athens' theatrical and cultural supremacy at a time when drama was proliferating all over the Hellenic world. By this time, the civic and social function of the theatre had changed, as had the aesthetics of the art form and the audiences it served. Yet the impetus to preserve theatrical works as texts in the fourth century has provided us with the means to examine how empathy played an important role in the socio-political life of the classical Athenian state.

Our earliest textual reference to Athenian drama is concerned with empathy. This is found in Herodotus, where he writes how the play *The Sack of Miletus* by Phyrnichus was banned and the playwright fined because the play so upset the audience who had known of the real and recent sack of Miletus by the Persians (6.21). We also catch a glimpse of theatrical empathy in action in Aristophanes' *Acharnians*. Here, Dicaeopolis goes to Euripides and begs him to borrow the costume of one of his characters, Telephus, so that he might earn the empathy of the assembly he must appeal to the following day (383–384). We can even find the seeds of dramatic empathy in the works of Homer. The disguised Odysseus is moved to tears by the songs of Demodocus, who tells of the wooden horse and the capture of Troy, and yet at the height of that moment where Odysseus is about to reveal his true identity, we are confronted with a striking image of the wife of a fallen enemy soldier, grieving over his body but then shoved into slavery by the harsh prod of the end of a spear shaft, "her face a mask of pain and suffering."[32] So many tragedies confronted the Athenian audience with similar predicaments and asked them to empathize with Trojan women, slaves, Furies, foreigners, and people who committed matricide, patricide, infanticide, and suicide. How were Athenian audiences, all of whom were required to serve in the military, expected to feel when watching the chorus of Theban women in Aeschylus's *Seven Against Thebes* sing of the horrors of *andrapodismos*, the systematic assaulting, sorting, separating, killing, and enslavement of people after a battle?[33] After all, they had participated in such actions immediately after the Persian invasion, according to Thucydides.[34] Was Aeschylus, a soldier himself, speaking out against such practices, or reminding his audience of what might happen to their own families if they did not defend the state? Either way, as my work with combat veterans of the wars in Vietnam, Iraq, and Afghanistan have taught me, any soldier who has experienced the effects of war on civilians cannot and almost certainly could not fail to be moved by such depictions.

Thucydides provides us with his view on such an event during the Peloponnesian War.[35] The former Athenian ally Mytilene had gone through an oligarchic coup, and the new government had attempted to go over to the Spartans. The Athenian forces had invaded and subjugated the island and were asking the Athenian assembly for instructions on how to proceed. During this debate, the demagogue Cleon pushes for the maximum penalty of complete *andrapodismos* to be exacted on the Mytileneans for their treachery. This would mean death for all men of military age, then the sorting of all non-combatants into those viable for sale as slaves and execution or abandonments for the elderly, infirm, and very young children. Families would be separated and the prisoners sold off to slave merchants. The city would then be destroyed or resettled with people who were loyal to Athens. The Athenian assembly initially voted to authorize their general in the field, Paches, to andrapodize the entire state, and the message was duly sent by ship.

The next day, many of the Athenians had a change of heart, believing that only the guilty should be punished, not the majority of the people who had been implicated in the rebellion by an oligarchic entity. A special assembly meeting was then called to decide the matter. At this meeting, Cleon chastised the Athenians, stating that "democracy is incapable of empire" because of their "change of mind" or "repentance" (3.37.1). Cleon then warned the assembly not to be persuaded by rhetoricians nor allow compassion to rule their judgment – they should "not be slaves to the pleasure of the ear" – and admonished them for acting more like the audience of a sophist than members of an assembly (3.38.7). For Cleon there were "three failings most fatal to empire – compassion, sentiment, and indulgence," and empathy (*eleos*) should only be given to those who are capable of feeling empathy in return (3.49.2–3). Cleon was unsuccessful in persuading a majority of Athenians, and the Mytilean civilians were famously spared in the nick of time. Though Cleon's opponent, Diodotus, had been careful not to use empathy in his rebuttal of Cleon's proposal, it is clear that it played an important role in deciding the fate of the people of Mytilene.

Around the same time, Euripides' produced his *Suppliants*. The play explores another aspect of the *Seven Against Thebes* myth and focuses on what happened after the Argive invaders had been defeated by the Thebans. A chorus made up of the mothers of the Argive dead come to Athens as suppliants to beg Theseus to force the Thebans to allow them to bury their sons. Initially, Theseus refuses to help, condemning the Argives for launching such a foolhardy and rash attack, but his mother, Aethra, after being moved by the powerful lament of the Argive women, persuades her son to act to preserve the customs of Greece and to not be called a coward in refusing the request of suppliants.

Theseus resolves to go to war over the rights of the women to bury their dead, and in another display of empathy a Theban messenger appeals to the Athenians for peace. This is an example of tragedy presenting alternate perspectives to the Athenian people and asking them to empathize with the victims of war. Here, the democratic act of voting for war is set against the empathetic image of death as the price of that war:

> Whenever war
> Comes up for the people's vote, no one counts on
> His own death; each thinks the other man
> Will suffer. But if death rose before your eyes
> When you cast your vote, Greece in its craze for spears
> Would not be destroyed in battle. All men know
> Which of two words is better: between peace and war,
> Which is evil and which is good, and how much more
> Peace benefits humankind. She is most dear
> To the Muses, hated by Vengeance. She loves

> Strong children, she rejoices in wealth. But we
> Choose war, in our evil, and enslave the weak,
> Man lording it over man, town over town.
>> Euripides, *Suppliants*, 470–482,
>> Tr. R. Warner & S. Scully

The final words of this study come from the final tragedy of the period we have been investigating, Euripides' *Bacchae*. The god returns to Thebes, the land of his birth, to force the city to accept his divinity. The old king Cadmus and the blind prophet Tiresias do so and enter dressed in fawn skins, eagerly dancing as best they can (200–209). Tiresias says that Bacchic revelry is a kind of madness that leads to "prophetic skill," as if the dissociative state of the Dionysian reveler leads to a kind of enlightenment. The chorus sings that Dionysos wants his followers to remove themselves from the normative existence and:

> To join the dances
> To laugh with the *aulos*
> To bring an end to thought.
>> Euripides, *Bacchae*, 380–383,
>> Tr. Paul Woodruff

I am proposing that tragedy offered its audiences a temporary "end to thought" in the form of dissonant experiences; they were dissociated from their normal processing fluency by strange settings and characters displaying abnormal behaviors; they watched their own young men play chorus characters who were distinctly "other," the masks otherworldly and eerily emotionally active, the movement communally understood yet heightened, and the music both familiar and strange. These dissociative and ambivalent aesthetic experiences led to audience absorption and a profound sense of empathy. This sense of mind-body dissonance and emotional ambivalence is encapsulated in Aristotle's famous statements about *catharsis* in the *Poetics*: that the vicarious pleasure derived from experiencing pity (*eleos*) and fear (*phobos*) is produced by the presentation of unexpected events (1453b). This is encapsulated in the final words of the chorus in *The Bacchae*:

> Many are the shapes the gods will take,
> Many the surprises they perform.
> What was thought likely did not transpire,
> And what was unlikely the god made easy.
> That is how this matter ended.
>> Euripides, *Bacchae*, 1388–1392,
>> Tr. Paul Woodruff

Coda: ancient empathy today

I subtitled this book *the imperative for theatre*, making the case within these pages that the affective experience of Greek drama contributed to the cognitive regime of the Athenian democracy and attempted to offer alternative views of prescient themes in contemporary life. I have argued that it offered a dissociative absorbing experience that increased empathy, not in the sense of "feeling with" the characters presented but in being emotionally provoked to project one's own feelings, which were perhaps repressed by the social norms of the prevailing culture, onto the masked actors before the *theatron*. This is the theatrical equivalent of Slaby's Interaction Theory of empathy, which proposes that instead of one interior mind transferring the affective state of another interior mind (Simulation Theory), there occurs mutual active co-engagement.[36] This is transmitted facially, bodily, and aurally by commonly understood affective markers, which are products of joint agency and active inferences and operations of participatory sense-making. The Greek theatre extended and amplified this co-engagement, not only between performer and audience but also among audience members seated in the collective "we-space" that was the open-air theatre. To adapt Slaby's model, this was a realm of *group* co-presence, of collective lived mutuality, bodily enacted among interacting individuals, who by inferring affective signals commonly understood within their culture, they could come to share powerful emotional states and collective feelings of empathy. Can theatre still do something similar today, and can Greek drama still prove effective as a powerful generator of empathy?

First, I want to briefly tackle the question of theatre today, which in general is a totally different experience from the open-air cult festival of fifth-century Athens. But it has something very important in common, and that is the presentation of *muthos* before an audience who have gathered in the same place to experience a live performance. This seems like a fact too basic to even articulate, but it really is the essence of a kind of shared empathetic experience that may be fast disappearing in our screen-mediatized culture. Here I run the risk of sounding like a Luddite, but, far from being encouraged to look up and out into the extra-personal, imagination-enhancing space of the sky, many of us now go about our business looking down at our handheld devices, barely noticing the environment we move through or the people we pass. Most of our mythic entertainment is delivered in private, in our homes or on those same devices, and the simple act of coming together in an audience to experience something challenging becomes rarer and rarer. We in the theatre have participated in this weakening of the embodied nature of our art form as we have been pushed to find new ways to finance our shows by seeking smaller theater spaces, casting stars from those screens to grace us with their ethereal presence, and hiking the cost of admission so that taking a risk or being challenged has to be tempered by the cost in time and money. Some artists try to break out of this by constantly exploring new ways to embody their art and reinvigorate "liveness-" environmental theatre, mixed-media presentations, alternative spaces, promenade productions, and shows in apartments, car, even phone booths. There still exists the craving to come together to experience the mimetic enchantment of the live actor before a living audience, as he or she brings us to understand the mental and emotional experiences that attend some predicament being performed – the very basis of Theory of Mind and the essence of empathy.

What theatre can still offer is the affective reception of another's view, a different perspective and empathy with another. The feeds on our smart phones and screens are not so smart when it comes to offering us a window on the world beyond our daily lives. Many of us fill these feeds with information that only reinforces what we already think, and the searchability of the computer seriously reduces the serendipitous discoveries of turning the page of a book or having our Default Mode Networks momentarily engaged by an arresting photo in a newspaper and leading us to a story we might otherwise never have read. Conversely, the evidence suggests that the presentation of alternate views in Greek drama had a profound impact on Athenian society, and we can certainly detect the impact those fifth-century plays had on the reputation of Athens as a city of culture, enlightenment, and thought.

What of Greek drama today? Can it still offer the kind of alternate experience that is powerful enough to change minds and even bring people together in empathy, perhaps even people who had once been enemies? Can we still understand and empathize with another? I draw on one recent example taken from public veteran's programs I have been involved with for the past several years. These programs run by Aquila Theatre since 2007 are called *Ancient Greeks/Modern Lives*, *You|Stories*, and *The Warrior Chorus*, and their aim has been to use ancient literature to create productive dialogue between members of the American veteran community and the public in an attempt to help Americans become more literate about the realities of war. This seems imperative in a society where less than 1% of the population serves in the military – a far cry from the total involvement of every male citizen in fifth-century Athens. In this work, vets read and perform Greek plays and take their themes as springboards for public discussions on often very difficult themes like combat trauma, suicide, treatment of the enemy, torture, military ethics, war and democracy, collective responsibility, homecoming, materialism, gender, violence, and politics. One of the program participants, a former Army Ranger who served multiple tours in Iraq and Afghanistan, described the use of ancient material in this way:

> I liked that the experiences were filtered through classical myth. This distance allows both performers and audience members to use their imaginations in an empathetic way, rather than merely evoking sympathy. This also helps free us from anachronistic terms such as PTSD or psychological wound, or whatever else they want to use to describe someone who has undergone a significant change due to military service. Classical myth places the emphasis back on character and story, and helps reject the laziness of labels. The abstract nature of myth also allows individuals to flesh out their own experiences with some combination of memory and imagination.[37]

In 2015, members of the Warrior Chorus visited Athens to perform their version of Sophocles' *Philoctetes* called *A Female Philoctetes*, which cast the lead role as a woman to focus on women in the military. The play was to be performed at the Cacoyannis Foundation Theatre, but the company arrived on the day the Greek banks were closed over the dispute concerning Greece's bailout by the Troika and its future in the European Union. The cafés were deserted (a true rarity in Athens), people were very worried, there were riot police on the street corners, and large protests were occurring. Greece had suffered mass unemployment, much of it in Athens, and on top of all of that was on the front lines of the growing refugee crisis, with people pouring in from Syria, Iraq, Afghanistan, and Pakistan. It was a crisis.

The show went ahead to a small, appreciative audience, but it was the next day where we were all surprised by the incredible power of drama to still create empathy. Aquila Theatre had partnered with a group in Athens that had been working with refugees on a version of Aeschylus's *Persians* entitled *We Are All Persians*. They had been using Aeschylus's play to create community, teach Greek, and explore the refugee experience through this work with classics. Now the two groups came together in a participatory workshop inspired by ancient drama, former American marines and soldiers alongside Afghans, Iraqis, and Syrians. One of the marines commented warily: "the last time I saw these guys I was kicking in their door and searching their house." Equally, the refugee members of the Athens group were very wary of meeting the American vets after seeing them perform the night before. Without a shared common language, the workshop involved movement around basic narrative themes inspired by images. In one an imposing marine was partnered with an actor and a young Pashtun man from Pakistan/Afghanistan. As the improvised scene progressed, the marine and the refugee moved from long stares to warily moving alongside each other, then they started to place their hands on each other and wrestle, struggling together, the larger marine seemed to get the upper hand and pinned the refugee down, but he did not surrender and kept resisting. Then all at once the marine ran into the wings and the refugee went in the other direction, both men clearly very emotional. Then they came together again and shook hands and that night ate together not as marines or refugees, Americans or Pashtuns, Muslims or Christians, but as a group of artists who had together shared their kinesthetic empathy and perhaps through that non-linguistic experience come to know something about each other's humanity. This is the imperative that still exists for theatre, perhaps more urgently now than ever before, and what I think the ancient Athenian dramatists were trying to achieve – a way in which to comprehend alternative viewpoints and to feel something for the predicaments of others. For Plato, the result of this affective empathetic performance culture was the *theatrocracy*, which emboldened the people to seek political liberty, to become fearless, and to resist authoritarian rule. The imperative of the theatre is still to be a mimetic mind that can embolden us to move beyond the world we think we know.

Notes

1 Vessel et al. 2011.
2 Starr 2013: 56.
3 Clark 2015: 166.
4 Xie et al. 2016.

5 Reeck et al. 2015.
6 DiGangi et al. 2016.
7 Mitchell-Boyask 2007.
8 Meineck 2016. See also Caston and Weineck 2016 and Meineck and Konstan 2014.
9 Sophocles, *Philocetes* 135.
10 Huang and Galinsky 2011.
11 Plato *Minos* 231a; Isocrates *Evagoras* 2.49 and 2.10; Aristotle, *Poetics* 1450b.16–21.
12 Rizzolatti et al. 2014.
13 Montgomery et al. 2016.
14 This example is inspired by my personal experiences working as a volunteer firefighter and emergency medical technician.
15 Topolinski and Strack 2010: 721.
16 Reber et al. 2004.
17 Klink and Wu 2014.
18 Alter and Oppenheimer 2006.
19 Topolinski and Strack 2010: 721. Also see Sievers et al. 2013.
20 Rees et al. 2013.
21 Vuoskoski et al. 2012: 311–317.
22 Aristophanes, *Wasps* 119–121.
23 Aristotle *Poetics 1449b 27*; *Politics* 1342a 10–15.
24 Starr 2013: 148.
25 Hein et al. 2016.
26 Cartledge 1997: 19.
27 Malhotra and Liyanage 2005.
28 Paluck 2009.
29 Paluck 2009: 582.
30 Plato, *Apology* 19c.
31 Xenophon, *Hellenica* 2.2.20.
32 Homer, *Odyssey* 8.521–531.
33 Meineck 2017.
34 Thucydides 1.98.2–4.
35 Thucydides 3.36–50.
36 Slaby 2014.
37 J. M. Meyer, from Public Witness Testimony Submitted to the Interior, Environment and Related Agencies Subcommittee Committee on Appropriations, U.S. House of Representatives Regarding FY 2015 Funding for the National Endowment for the Humanities by Dr. Peter Meineck, New York University and Aquila Theatre (April 10, 2014).

Bibliography

Alter, A.L. and Oppenheimer, D.M. 2006. "Predicting short-term stock fluctuations by using processing fluency" in *Proceedings of the National Academy of Sciences* 103.24: 9369–9372.

Cartledge, P. 1997. "'Deep plays': Theatre as process in Greek civic life" in Easterling, P.E. (Ed.), *The Cambridge companion to Greek tragedy*. Cambridge: Cambridge University Press: 3–35.

Caston, V. and Weineck, S.M. 2016. *Our ancient wars: Rethinking war through the classics*. Ann Arbor, MI: University of Michigan Press.

Clark, A. 2015. *Surfing uncertainty: Prediction, action, and the embodied mind*. Oxford: Oxford University Press.

DiGangi, J.A., Tadayyon, A., Fitzgerald, D.A., Rabinak, C.A., Kennedy, A., Klumpp, H., Rauch, S.A. and Phan, K.L. 2016. "Reduced default mode network connectivity following combat trauma" in *Neuroscience Letters* 615: 37–43.

Hein, G., Engelmann, J.B., Vollberg, M.C. and Tobler, P.N. 2016. "How learning shapes the empathic brain" in *Proceedings of the National Academy of Sciences* 113.1: 80–85.

Huang, L. and Galinsky, A.D. 2011. "Mind – body dissonance conflict between the senses expands the mind's horizons" in *Social Psychological and Personality Science* 2.4: 351–359.

Klink, R.R. and Wu, L. 2014. "The role of position, type, and combination of sound symbolism imbeds in brand names" in *Marketing Letters* 25.1: 13–24.

Malhotra, D. and Liyanage, S. 2005. "Long-term effects of peace workshops in protracted conflicts" in *Journal of Conflict Resolution* 49.6: 908–924.

Meineck, P. 2016. "Combat trauma and the tragic stage" in Caston, V. and Weineck, S.M. (Eds.), *Our ancient wars: Rethinking war through the classics*. Ann Arbor, MI: University of Michigan Press: 184–210.

Meineck, P. 2017. "Thebes as high collateral damage target: Moral accountability for killing in *Seven Against Thebes*" in Torrance, E. (Ed.), *Aeschylus and war: Comparative perspectives on Seven Against Thebes*. London: Routledge: 49–70.

Meineck, P. and Konstan, D. (Eds.). 2014. *Combat trauma and the ancient Greeks*. New York: Palgrave Macmillan.

Mitchell-Boyask, R. 2007. *Plague and the Athenian imagination: Drama, history, and the cult of Asclepius*. Cambridge: Cambridge University Press.

Montgomery, C.B., Allison, C., Lai, M.C., Cassidy, S., Langdon, P.E. and Baron-Cohen, S. 2016. "Do adults with high functioning autism or Asperger Syndrome differ in empathy and emotion recognition?" in *Journal of Autism and Developmental Disorders* 46.6: 1931–1940.

Paluck, E.L. 2009. "Reducing intergroup prejudice and conflict using the media: A field experiment in Rwanda" in *Journal of Personality and Social Psychology* 96.3: 574–587.

Reber, R., Schwarz, N. and Winkielman, P. 2004. "Processing fluency and aesthetic pleasure: Is beauty in the perceiver's processing experience?" in *Personality and Social Psychology Review* 8.4: 364–382.

Reeck, C., Ames, D.R. and Ochsner, K.N. 2015. "The social regulation of emotion: An integrative, cross-disciplinary model" in *Trends Cognitive: Science* 20: 47–63.

Rees, L., Rothman, N.B., Lehavy, R. and Sanchez-Burks, J. 2013. "The ambivalent mind can be a wise mind: Emotional ambivalence increases judgment accuracy" in *Journal of Experimental Social Psychology* 49.3: 360–367.

Rizzolatti, G., Cattaneo, L., Fabbri-Destro, M. and Rozzi, S. 2014. "Cortical mechanisms underlying the organization of goal-directed actions and mirror neuron-based action understanding" in *Physiological Reviews* 94.2: 655–706.

Sievers, B., Polansky, L., Casey, M. and Wheatley, T. 2013. "Music and movement share a dynamic structure that supports universal expressions of emotion" in *Proceedings of the National Academy of Sciences of the United States of America* 110.1: 70–75.

Slaby, J. 2014. "Empathy's blind spot" in *Medicine, Health Care and Philosophy* 17.2: 249–258.

Starr, G.G. 2013. *Feeling beauty: The neuroscience of Aesthetic experience*. Cambridge, MA: MIT Press.

Topolinski, S. and Strack, F. 2010. "False fame prevented: Avoiding fluency effects without judgmental correction" in *Journal of Personality and Social Psychology* 98.5: 721–733.

Vessel, E.A., Starr, G.G. and Rubin, N. 2011. "The brain on art: Intense aesthetic experience activates the default mode network" in *Frontiers in Human Neuroscience* 6.66: 1–17.

Vuoskoski, J.K., Thompson, W.F., McIlwain, D. and Eerola, T. 2012. "Who enjoys listening to sad music and why?" in *Music Perception: An Interdisciplinary Journal* 29.3: 311–317.

Xie, X., Bratec, S.M., Schmid, G., Meng, C., Doll, A., Wohlschläger, A., Finke, K., Förstl, H., Zimmer, C., Pekrun, R. and Schilbach, L. 2016. "How do you make me feel better? Social cognitive emotion regulation and the default mode network" in *NeuroImage* 134: 270–280.

Index